Test Bank

for

Physics
A World View

Sixth Edition

Larry D. Kirkpatrick
Montana State University

Gregory E. Francis
Montana State University

THOMSON

BROOKS/COLE

Australia • Brazil • Canada • Mexico • Singapore • Spain • United Kingdom • United States

Thomson Higher Education
10 Davis Drive
Belmont, CA 94002-3098
USA

For more information about our products, contact us at:
Thomson Learning Academic Resource Center
1-800-423-0563

For permission to use material from this text or product, submit a request online at
http://www.thomsonrights.com.
Any additional questions about permissions can be submitted by email to **thomsonrights@thomson.com.**

TEST BANK

PHYSICS: A World View

Table of Contents

Chapter 1: A World View

1. Let's start off with an easy one. The subject matter in this course is
   a. psychology.
   * b. physics.
   c. philosophy.
   d. physiology.

2. I am very happy that the first question is an easy one to get me off to a good start in this course. The subject matter of this course is
   a. physical education.
   b. psychiatry.
   c. philosophy.
   * d. physics.

3. I am very happy that the first question is an easy one to get me off to a good start in this course. The instructor for this course is
   * a. Kirkpatrick.
   b. Newton.
   c. Galileo.
   d. Aristotle.

4. Let's begin with a question that everyone should get correct. The subject matter of this course is
   a. psychics.
   b. fysix.
   * c. physics.
   d. phonics.

5. The recognition of the permanence of the physical size of objects is
   a. a concept with which we are born
   * b. a learned invariant
   c. obvious from the constancy of the retinal image

6. Which of the following is not a valid criterion for the acceptance of a physical law or theory?
   a. agreement with past data
   b. based on scientific principles
   c. ability to predict future results
   * d. prestige of the scientist proposing it

Chapter 1   A World View

7. In physics a world view is
      a. the visual stimuli that allow the physicist to experience the microscopic world.
      b. the idea that the Sun is at the center of the Solar System.
    * c. a shared set of ideas that represents the current explanations of how the material world operates.
      d. the unchanging rules used to predict the behavior of physical systems.

8. In the physics world view an invariant is
      a. the agreement of theory and experiment.
    * b. something that does not change.
      c. a belief that is based on prejudice.
      d. a physical law that did not survive experimental tests.

9. An operational definition is a
    * a. rule for attaching a number to a concept.
      b. logical argument for obtaining a law of physics.
      c. method of excising the truth from nature.
      d. medical procedure performed on a physics teacher.

10. Bode's law giving the sizes of the orbits of the planets is no longer considered to be a physical law because it
      a. did not test for the data known at the time it was proposed
      b. did not make any predictions that could be tested
      c. was actually proposed by the Titus of Wittenburg
    * d. it made predictions that were incorrect

11. Bode's law giving the sizes of the orbits of the planets is no longer considered to be a physical law because it
      a. did not agree with the data known at the time it was proposed.
      b. did not make any predictions that could be tested.
      c. was actually proposed by the Titus of Wittenburg.
    * d. was not based on physical principles.

12. Which of the following is not one of the branches of physics?
      a. mechanics
      b. atomic
      c. nuclear
    * d. astrology

13. Which of the following lists of major countries shows those that have not accepted the metric system of units?
    * a. only the United States
      b. United States, Saudi Arabia, Iran, Iraq
      c. United States, Russia, Ukraine, Belarus
      d. United States, Denmark, Finland, Norway, Sweden

14. Besides the United States, which of the following major countries has not adopted the metric system of units?
    * a. No other
      b. Great Britain
      c. Australia
      d. Russia

15. How many millimeters are there in one kilometer?
      a. 10
      b. 100
      c. 1000
    * d. 1,000,000

16. A meter is about the same length as
      a. a mile.
      b. a foot.
    * c. a yard.
      d. an inch.

17. Approximately how long is a midsized car?
      a. 2 meters
      b. 3 meters
    * c. 4 meters
      d. 5 meters

18. A kilometer is about the same length as
      a. a football field.
    * b. 5/8 mile.
      c. one mile.
      d. 10 miles.

Chapter 1   A World View

19. The length of the desk in the front of the classroom is approximately _____ long.
      a. 15 mm
      b. 15 cm
  * c. 1.5 m
      d. 0.15 km

20. The height of the ceiling in this room is approximately
  * a. 3 m.
      b. 30 cm.
      c. 300 mm.
      d. 0.03 km.

21. The height of the ceiling in this classroom is about
      a. 1 meter.
      b. 2 meters.
  * c. 3 meters.
      d. 5 meters.

22. Which of the following is equal to 86,400 seconds?
      a. 8.64×10^3 seconds
  * b. 8.64×10^4 seconds
      c. 8.64×10^5 seconds
      d. 8.64×10^6 seconds

23. The number of seconds in a week is equal to 6.05×10^5 seconds. This can also we written as
      a. 60,500 seconds.
      b. 605,000 seconds.
  * c. 6,050,000 seconds.
      d. 60,500,000 seconds.

24. A mass of 4.3×10^3 grams can also be written as
      a. 43 grams.
      b. 430 grams.
  * c. 4300 grams.
      d. 4,300 grams.

25. The diameter of a typical atom is approximately 0.000 000 000 1 meter. This can also be written as
      a. 1×10^{-9} meter.
  * b. 1×10^{-10} meter.
      c. 1×10^{-11} meter.
      d. 1×10^{10} meter.

26. What is the product of 4×10^4 and 2×10^8?
       a. 8×10^4
       b. 8×10^8
   * c. 8×10^{12}
       d. 8×10^{32}

27. What is the result of dividing 6×10^8 by 2×10^4 ?
   * a. 3×10^4
       b. 3×10^8
       c. 3×10^{12}
       d. 3×10^2

28. When you calculate the speed (in meters per second) in an experiment, your calculator display reads 12.666667. If you are asked to record your result to three significant figures, you should write
       a. 12.6 m/s.
   * b. 12.7 m/s.
       c. 12.666 m/s.
       d. 12.667 m/s.

29. How many significant figures are there in the measurement 0.000 234 m^2 ?
   * a. 3
       b. 4
       c. 5
       d. 6

30. Which of the following expressions gives the number of seconds in one week?

   * a. $\left(\dfrac{7 \text{ days}}{1 \text{ week}}\right)\left(\dfrac{24 \text{ h}}{1 \text{ day}}\right)\left(\dfrac{3600 \text{ s}}{1 \text{ h}}\right)$

       b. $\left(\dfrac{7 \text{ days}}{1 \text{ week}}\right)\left(\dfrac{1 \text{ day}}{24 \text{ h}}\right)\left(\dfrac{1 \text{ h}}{3600 \text{ s}}\right)$

       c. $\left(\dfrac{1 \text{ week}}{7 \text{ days}}\right)\left(\dfrac{24 \text{ h}}{1 \text{ day}}\right)\left(\dfrac{3600 \text{ s}}{1 \text{ h}}\right)$

       d. $\left(\dfrac{1 \text{ week}}{7 \text{ days}}\right)\left(\dfrac{1 \text{ day}}{24 \text{ h}}\right)\left(\dfrac{1 \text{ h}}{3600 \text{ s}}\right)$

31. Given that a mile is equal to 1600 m, which calculation shows the correct conversion of a speed of 25 mph to the same speed in m/s?

  *   a. $\left(25\ \dfrac{\text{miles}}{\text{h}}\right)\left[\dfrac{1\ \text{h}}{3600\ \text{s}}\right]\left[\dfrac{1600\ \text{m}}{1\ \text{mile}}\right]$

      b. $\left(25\ \dfrac{\text{miles}}{\text{h}}\right)\left[\dfrac{1\ \text{h}}{3600\ \text{s}}\right]\left[\dfrac{1\ \text{mile}}{1600\ \text{m}}\right]$

      c. $\left(25\ \dfrac{\text{miles}}{\text{h}}\right)\left[\dfrac{3600\ \text{s}}{1\ \text{h}}\right]\left[\dfrac{1\ \text{mile}}{1600\ \text{m}}\right]$

      d. $\left(25\ \dfrac{\text{miles}}{\text{h}}\right)\left[\dfrac{3600\ \text{s}}{1\ \text{h}}\right]\left[\dfrac{1600\ \text{m}}{1\ \text{mile}}\right]$

32. Given that a mile is equal to 1600 m, which calculation shows the correct conversion of a speed of 20 m/s to the same speed in mph?

      a. $\left(20\ \dfrac{\text{m}}{\text{s}}\right)\left[\dfrac{1\ \text{h}}{3600\ \text{s}}\right]\left[\dfrac{1600\ \text{m}}{1\ \text{mile}}\right]$

      b. $\left(20\ \dfrac{\text{m}}{\text{s}}\right)\left[\dfrac{1\ \text{h}}{3600\ \text{s}}\right]\left[\dfrac{1\ \text{mile}}{1600\ \text{m}}\right]$

  *   c. $\left(20\ \dfrac{\text{m}}{\text{s}}\right)\left[\dfrac{3600\ \text{s}}{1\ \text{h}}\right]\left[\dfrac{1\ \text{mile}}{1600\ \text{m}}\right]$

      d. $\left(20\ \dfrac{\text{m}}{\text{s}}\right)\left[\dfrac{3600\ \text{s}}{1\ \text{h}}\right]\left[\dfrac{1600\ \text{m}}{1\ \text{mile}}\right]$

33. If there are 8 furlongs in one mile, how fast in miles per hour can a horse run if it can complete a 12-furlong race in 2 minutes?

      a) $\left(\dfrac{12\ \text{furlong}}{2\ \text{min}}\right)\left[\dfrac{8\ \text{mile}}{1\ \text{furlong}}\right]\left[\dfrac{60\ \text{min}}{1\ \text{h}}\right]$

  *   b) $\left(\dfrac{12\ \text{furlong}}{2\ \text{min}}\right)\left[\dfrac{1\ \text{mile}}{8\ \text{furlong}}\right]\left[\dfrac{60\ \text{min}}{1\ \text{h}}\right]$

      c) $\left(\dfrac{12\ \text{furlong}}{2\ \text{min}}\right)\left[\dfrac{8\ \text{furlong}}{1\ \text{mile}}\right]\left[\dfrac{60\ \text{min}}{1\ \text{h}}\right]$

      d) $\left(\dfrac{12\ \text{furlong}}{2\ \text{min}}\right)\left[\dfrac{1\ \text{mile}}{8\ \text{furlong}}\right]\left[\dfrac{30\ \text{min}}{1\ \text{h}}\right]$

Physics: A World View, Sixth Edition **by Larry Kirkpatrick and Gregory Francis**

34. A certain piece of fabric sells in Canada for $4.75 CD per meter. Given that 1 meter = 1.09 yards and $1.00 US= $0.65 CD, what is the price of the fabric in US dollars per yard?

a. $\left(\dfrac{\$4.75\text{ CD}}{1\text{ m}}\right)\left[\dfrac{\$1.00\text{ US}}{\$0.65\text{ CD}}\right]\left[\dfrac{1.09\text{ yard}}{1\text{ m}}\right]$

b. $\left(\dfrac{\$4.75\text{ CD}}{1\text{ m}}\right)\left[\dfrac{\$0.65\text{ CD}}{\$1.00\text{ US}}\right]\left[\dfrac{1\text{ m}}{1.09\text{ yard}}\right]$

* c. $\left(\dfrac{\$4.75\text{ CD}}{1\text{ m}}\right)\left[\dfrac{\$1.00\text{ US}}{\$0.65\text{ CD}}\right]\left[\dfrac{1\text{ m}}{1.09\text{ yard}}\right]$

d. $\left(\dfrac{\$4.75\text{ CD}}{1\text{ m}}\right)\left[\dfrac{\$0.65\text{ CD}}{\$1.00\text{ US}}\right]\left[\dfrac{1.09\text{ yard}}{1\text{ m}}\right]$

35. On average, Americans eat 28 pounds of bananas per year. On average, how many pounds of bananas are eaten each day in the U.S.?

a. $\left(\dfrac{28\text{ lb}}{\text{person}\cdot\text{y}}\right)\left[\dfrac{1}{290\text{ million people}}\right]\left[\dfrac{365\text{ days}}{1\text{ y}}\right]$

b. $\left(\dfrac{28\text{ lb}}{\text{person}\cdot\text{y}}\right)\left[\dfrac{1}{290\text{ million people}}\right]\left[\dfrac{1\text{ y}}{365\text{ days}}\right]$

c. $\left(\dfrac{28\text{ lb}}{\text{person}\cdot\text{y}}\right)(290\text{ million people})\left[\dfrac{365\text{ days}}{1\text{ y}}\right]$

* d. $\left(\dfrac{28\text{ lb}}{\text{person}\cdot\text{y}}\right)(290\text{ million people})\left[\dfrac{1\text{ y}}{365\text{ days}}\right]$

36. If a league is equal to 3 miles, how many miles are there in "20,000 Leagues Under the Sea?"
    a. 6,667
    b. 19,997
    c. 20,003
* d. 60,000

37. A speed of 90 km/hr is equal to _____ m/s.
* a. 25
    b. 90
    c. 150
    d. 324

38. A speed of 36 m/s is equal to _____ km/hr.
    a. 6
    b. 10
    c. 60
    * d. 130

39. A speed of 3 m/s is approximately equal to _____ km/hr.
    a. 0.2
    b. 0.8
    * c. 11
    d. 50

40. If there are 8 furlongs in one mile, how many furlongs does a horse run in a 4-mile race?
    a. 2
    b. 4
    c. 8
    * d. 32

41. If there are 8 furlongs in one mile, how fast in miles per hour can a horse run if it can complete a 12-furlong race in 2 minutes?
    a. 15 mph
    b. 30 mph
    * c. 45 mph
    d. 60 mph

Chapter 2: Describing Motion

1. The average speed of an object is defined to be the
   * a. distance it travels divided by the time it takes.
     b. distance it travels multiplied by the time it takes.
     c. change in its velocity divided by the time it takes.
     d. change in its velocity multiplied by the time it takes.

2. A runner in the Boston marathon covered the first 20 miles in a time of 4 hours. How fast was he running when he passed the 15 mile marker?
     a. 5 mph
     b. 10 mph
     c. 20 mph
   * d. We don't know.

3. A Honda Civic travels from milepost 405 to milepost 455 between 1:00 and 2:00 pm. A Subaru Legacy travels from milepost 200 to milepost 240 during the same time. Which car was traveling faster at 1:30 pm?
     a. Honda
     b. Subaru
     c. They were traveling at the same speed.
   * d. There is not enough information to be able to say.

4. A train covers 60 miles between 2 pm and 4 pm. How fast was it traveling at 3 p.m.?
     a. 15 mph
     b. 30 mph
     c. 60 mph
   * d. Not enough information is given to be able to say.

5. Car A travels from milepost 343 to milepost 349 in 5 minutes. Car B travels from milepost 493 to milepost 499 in 5 minutes. Which car has the greater average speed?
     a. Car A
     b. Car B
   * c. Their average speeds are the same.
     d. There is not enough information to be able to say.

6. A yellow car takes 10 minutes to go from milepost 101 to milepost 109. A red car takes 10 minutes to go from milepost 11 to milepost 21. Which car has the higher average speed?
     a. the yellow one
   * b. the red one
     c. Their average speeds are the same.
     d. Not enough information is given to be able to say.

Chapter 2   Describing Motion

7. In Aesop's fable of the tortoise and the hare, the "faster" hare loses the race to the slow and steady tortoise. During the race, which animal has the greater average speed?
    *   a. the tortoise
         b. the hare
         c. Both have the same average speed.
         d. There is not enough information to say.

8. Pat and Chris both travel from Los Angeles to New York along the same route. Pat rides a bicycle while Chris drives a fancy sports car. Unfortunately, Chris's car breaks down in Phoenix for over a week, causing the two to arrive in New York at exactly the same time. Which statement is true?
    *   a. Pat and Chris had the same average speed.
         b. Chris had the higher average speed.
         c. Pat had the higher average speed.

9. On a trip to Helena, you stop for a 15-minute coffee break in Three Forks and arrive in Helena two hours after leaving Bozeman. If you assume that it is 100 miles to Helena, your average speed would be 50 mph. Which of the following statements about this trip is correct?
         a. To average 50 mph the car must have gone 100 mph for 15 minutes of the trip.
         b. The average speed is not 50 mph but what was indicated on the speedometer.
         c. You cannot average 50 mph if the speed is zero for any part of the trip.
    *   d. The car must have traveled faster than 50 mph for part of the trip.

10. A cruise ship covers a distance of 80 miles during the watch that lasts from midnight to 8 am. How fast was the ship going at 4 am if the speed of the ship was constant during the watch?
         a. 80 miles/hour
         b. 80 miles
    *   c. 10 miles/hour
         d. We don't have enough information to be able to say.

11. If a woman walks at a speed of 2 miles/hour for 3 hours, she will have walked
         a. 2 miles.
         b. 5 miles.
    *   c. 6 miles.
         d. 9 miles.

12. If a marathoner can run with an average speed of 10 mph, how far could she run in 2 hours?
         a. 5 miles
         b. 10 miles
         c. 12 miles
    *   d. 20 miles

13. How many hours are required to make a 4400-km trip across the United States if you average 80 km/h?
     a. 45 h
     b. 50 h
  * c. 55 h
     d. 60 h

14. Approximately how fast can a person run?
     a. 1 m/s
  * b. 10 m/s
     c. 100 m/min
     d. 1 km/h

15. The instantaneous speed of an object is defined to be the
     a. distance it travels divided by the time it takes.
     b. distance it travels multiplied by the time it takes.
  * c. average speed determined over an infinitesimally small time interval.
     d. value of the average speed at the midpoint of the time interval.

16. Which of the following could be a velocity?
     a. 5 meters west
     b. 5 meters per second
  * c. 5 meters per second west
     d. 5 meters per second per second

17. The average acceleration of an object is defined to be the
     a. distance it travels divided by the time it takes.
  * b. change in its velocity divided by the time it takes.
     c. change in its speed divided by the time it takes.
     d. average of the accelerations during the two halves of the trip.

18. An object is accelerating
     a. only when its speed changes.
     b. only when its direction changes.
  * c. when its speed or direction changes.

19. Which of the following is considered to be an "accelerator" in an automobile?
     a. brake pedal
     b. gas pedal
     c. steering wheel
  * d. All are accelerators.

20. If a car changes speed from 60 mph to 66 mph in 1 minute, its average acceleration is
      a. 60 mph/minute.
      b. 63 mph/minute.
      c. 66 mph/minute.
   * d. 6 mph/minute.

21. If a car changes speed from 40 mph to 48 mph in 2 minutes, its average acceleration is
      a. 4 miles/hour.
      b. 44 miles/hour/minute.
      c. 8 miles/hour/minute.
   * d. 4 miles/hour/minute.

22. If a car requires 10 seconds to accelerate from zero to 60 mph, its average acceleration is
      a. 600 mph/second.
      b. 60 mph/second.
      c. 10 mph/second.
   * d. 6 mph/second.

23. A Chevrolet Corvette can accelerate from 0 to 60 mph in 5.2 s. What is the car's average acceleration?
      a. 0 mph/s
   * b. 11.5 mph/s
      c. 9.8 mph/s
      d. 312 mph/s

24. A pitcher requires about 0.1 second to throw a baseball. If the ball leaves his hand with a speed of 40 m/s, what is the ball's average acceleration?
      a. 40 m/s^2
   * b. 400 m/s^2
      c. 400 m/s
      d. 400 m

25. A car is accelerating at 4 m/s^2. At some time it is traveling at 40 m/s. How fast will it be going 1 s later?
      a. 36 m/s
      b. 40 m/s
      c. 44 m/s
   * d. 36 m/s or 44 m/s

26. A child traveling 5 m/s on a sled passes her younger brother. If her average acceleration on the sledding hill is 2 m/s^2, how fast is she traveling when she passes her older brother 4 s later?

    a. 7 m/s
    b. 8 m/s
    c. 10 m/s
*   d. 13 m/s

27. In the strobe diagram below the ball is moving from left to right. Which statement best describes the motion? The ball is

  o   o   o   o   o o o

    a. moving with a constant speed.
    b. speeding up.
*   c. slowing down.
    d. not accelerating.

28. Which of the following strobe diagrams corresponds to the situation where a ball rolls from left to right and continually speeds up?

    a. o   o   o   o   o   o   o
*   b. o o   o   o   o
    c. o o   o o o   o   o   o o
    d. o   o   o   o   o o o

29. Which of the following strobe diagrams corresponds to the situation where a ball rolls from right to left and continually slows down?

    a. o   o   o   o   o   o   o
*   b. o o   o   o   o
    c. o o   o o o   o   o   o o
    d. o   o   o   o   o o o

30. Which of the following strobe diagrams does NOT correspond to a situation where the ball is accelerating?

*   a. o   o   o   o   o   o   o
    b. o o   o   o   o
    c. o o   o o o   o   o   o o
    d. o   o   o   o   o o o

Chapter 2   Describing Motion

31. Which statement best describes the motion of the ball shown in the strobe diagram below? (Assume the ball moves from left to right.) The ball is

    O . . . . O . . . . O . . . . O . . . . O . . . . O

    * a. moving with constant speed.
      b. speeding up.
      c. accelerating.
      d. slowing down.

32. Which statement best describes the motion of the ball shown in the strobe diagram below? (Assume the ball moves from left to right.) The ball is

    O . . . . O . . . . O . . . . O . . . . . . O . . . . O . . OO

      a. moving with constant speed.
      b. speeding up.
    * c. accelerating.
      d. stopped

33. When we say that light objects and heavy objects fall at the same rate, what assumption are we making?

      a. They have the same shape.
    * b. They are falling in a vacuum.
      c. They are made of the same material.
      d. They have the same size.

34. A ping-pong ball and a golf ball have approximately the same size but very different masses. Which hits the ground first if you drop them simultaneously while standing on the Moon?
      a. the ping-pong ball
      b. the golf ball
    * c. They hit simultaneously.
      d. We are not able to predict the results.

35. The Moon is a nice place to study free-fall because it has no atmosphere. If an astronaut on the Moon simultaneously drops a hammer and a feather from the same height, which one hits the ground first?
      a. the hammer
      b. the feather
    * c. They hit at the same time.
      d. They don't fall.

36. We claimed that if the air resistance could be neglected, all objects on the Moon would fall at
     a. the same constant speed.
     b. an increasing acceleration.
   * c. the same constant acceleration.
     d. a decreasing acceleration.

37. A sheet of paper and a book fell at different rates in the classroom until the paper was wadded up into a ball. We then claimed that if the air resistance could be neglected, all objects would fall at
     a. different constant speeds depending on the type of material.
     b. the same constant speed regardless of the type of material.
     c. the same constant speed regardless of how much they weigh.
   * d. the same constant acceleration.

38. A student decides to test Aristotle's and Galileo's ideas about free-fall by simultaneously dropping a 20-lb. ball and a 1-lb. ball from the top of a grain elevator. The two balls have the same size and shape. What actually happens? (Do not neglect air resistance!).
   * a. The 20-lb. ball hits first.
     b. The 1-lb. ball hits first.
     c. They hit simultaneously.
     d. We are not able to predict the results.

39. A ping-pong ball and a golf ball have approximately the same size but very different masses. Which hits the ground first if you drop them simultaneously from a tall building? Do not ignore the effects of the air.
     a. the ping-pong ball
   * b. the golf ball
     c. They hit simultaneously.
     d. We are not able to predict the results.

40. If we do *not* neglect air resistance, during which, if any, of the first 5 s of free fall does a ball's speed change the most?
   * a. first second
     b. third second
     c. fifth second
     d. The speed changes the same amount each second.

41. If we ignore air resistance, the acceleration of an object that is falling downward is constant. How do you suppose the acceleration would change if we do *not* ignore air resistance?
     a. The acceleration increases.
     b. The acceleration does not change.
   * c. The acceleration decreases.

Chapter 2   Describing Motion

42. A ball is thrown straight up into the air. If we do not ignore air resistance, the acceleration of the
    ball as it is traveling upward is
         a. 9.8 m/s^2.
    * b. greater than 9.8 m/s^2.
         c. less than 9.8 m/s^2.
         d. zero.

43. A ball is dropped in air. If we do not ignore air resistance, the acceleration of the ball is
         a. 9.8 m/s^2.
         b. greater than 9.8 m/s^2.
    * c. less than 9.8 m/s^2.
         d. zero.

44. If the mass of an object in free fall is doubled, its acceleration
         a. doubles.
         b. increases by a factor of four.
    * c. stays the same.
         d. is cut in half.

45. The motion of a block sliding down a frictionless ramp can be described as motion with
         a. a constant speed.
         b. a constant acceleration greater than 10 m/s/s.
    * c. a constant acceleration less than 10 m/s/s.
         d. a constant speed that depends on the steepness of the ramp.

46. The motion of a ball or cylinder rolling down a ramp is one with
         a. constant speed.
         b. increasing acceleration.
    * c. constant acceleration.
         d. decreasing acceleration.

47. You are bouncing on a trampoline while holding a bowling ball. As your feet leave the
    trampoline, you let go of the bowling ball. You will rise _____ than you would have if you
    had held onto the bowling ball.
         a. higher
    * b. to the same height
         c. lower

Physics: A World View, Sixth Edition by Larry Kirkpatrick and Gregory Francis

48. You are bouncing on a trampoline while holding a bowling ball. As your feet leave the trampoline, you let go of the bowling ball. When you reach your maximum height, the bowling ball is
  a. above you.
 * b. beside you.
  c. below you.

49. Suppose that you look out a tenth-floor window and see a ball falling at 5 m/s. How fast will this ball be falling 1 s later?
  a. 5 m/s
  b. 10 m/s
 * c. 15 m/s
  d. 20 m/s

50. Suppose that you look out a tenth-floor window and see a ball falling at 5 m/s. How fast will this ball be falling 2 s later?
  a. 5 m/s
  b. 15 m/s
 * c. 25 m/s
  d. 35 m/s

51. An object is dropped off a cliff. What is its instantaneous speed 3 s later?
  a. 15 m/s
 * b. 30 m/s
  c. 45 m/s
  d. 60 m/s

52. If a ball is dropped from rest, it will fall 5 m during the first second. How far will it fall during the first 2 s?
  a. 10 m
  b. 15 m
 * c. 20 m
  d. 25 m

53. If a ball is dropped from rest, it will fall 5 m during the first second. How far will it fall during the second second?
  a. 5 m
  b. 10 m
 * c. 15 m
  d. 20 m

Chapter 2   Describing Motion

54. An object is dropped off a cliff. How far will the object fall during the next 4 s?
    - a. 20 m
    - b. 45 m
    - * c. 80 m
    - d. 125 m

55. You decide to launch a ball vertically so that a friend located 45 m above you can catch it. What is the minimum launch speed you can use?
    - a. 4.5 m/s
    - b. 20 m/s
    - * c. 30 m/s
    - d. 45 m/s

56. A ball is thrown vertically upward and you know that its speed is 20 m/s as it leaves the thrower's hand. What is the speed of the ball 1 s later?
    - a. 30 m/s
    - b. 20 m/s
    - * c. 10 m/s
    - d. zero

57. A rock is thrown vertically upward with a speed of 15 m/s. What are its speed and direction 2 s later?
    - a. 10 m/s upward
    - b. 5 m/s upward
    - c. zero
    - * d. 5 m/s downward

58. A golf ball is thrown vertically upward with a speed of 30 m/s. How long does it take to get to the top of its path?
    - a. 1 s
    - b. 2 s
    - * c. 3 s
    - d. 4 s

59. You throw a ball straight up at 30 m/s. How many seconds elapse before it is traveling downward at 10 m/s?
    - a. 2 s
    - b. 3 s
    - * c. 4 s
    - d. 5 s

Physics: A World View, Sixth Edition **by Larry Kirkpatrick and Gregory Francis**

60. If we use plus and minus signs to indicate the directions of velocity and acceleration, in which of the following situations does the object speed up?
    a. positive velocity and negative acceleration
    b. negative velocity and positive acceleration
    c. positive velocity and zero acceleration
*   d. negative velocity and negative acceleration

61. A car traveling westward at 20 m/s turns around and travels eastward at 5 m/s. What is the change in velocity of the car?
    a. 15 m/s west
    b. 25 m/s
    c. 25 m/s west
*   d. 25 m/s east

62. A car traveling westward at 20 m/s turns around and travels eastward at 15 m/s. If this takes place in 5 s, what is the average acceleration of the car?
    a. 1 m/s^2 west
    b. 7 m/s^2
    c. 7 m/s^2 west
*   d. 7 m/s^2 east

63. A car initially traveling westward at 20 m/s has a constant acceleration of 1 m/s^2 westward. How far does the car travel in the first 10 s?
    a. 200 m
    b. 210 m
*   c. 250 m
    d. 300 m

64. A car initially traveling westward at 20 m/s has a constant acceleration of 1 m/s^2 eastward. How far does the car travel in the first 10 s?
*   a. 150 m
    b. 190 m
    c. 200 m
    d. 250 m

Chapter 3: Explaining Motion

1. The property of an object at rest to remain at rest is known as
   - a. inertness.
   - * b. inertia.
   - c. resistance.
   - d. sluggishness.

2. If there is no net force acting on an object, its motion will be one with _____ acceleration.
   - * a. zero
   - b. a constant, non-zero
   - c. an increasing
   - d. a decreasing

3. Assume that you are driving down a straight road at constant speed. A small ball is tied on the end of a string hanging from the rear view mirror. Which way will the ball swing when you apply the brakes?
   - * a. forward
   - b. backward
   - c. It will not swing.

4. If an object moves in a straight line with a constant speed, we can conclude that
   - a. the object has inertia.
   - b. there are no forces acting on the object.
   - c. there must be at least two forces acting on the object.
   - * d. there is no unbalanced force acting on the object.

5. If an object moves with a constant velocity, we can conclude that
   - a. it is moving toward its natural place.
   - b. there are no forces acting on it.
   - * c. there is no unbalanced (net) force acting on it.
   - d. it has a very large inertia.

6. What is the net force on an 900-kg airplane flying with a constant velocity of 180 km/hour north?
   - * a. zero
   - b. 180 N
   - c. 900 N
   - d. 162,000 N

7. What is the net force on a 2000-kg car that is traveling in a straight line at a constant speed of 40 m/s?
   * a. zero
     b. 40 N
     c. 2000 N
     d. (40 m/s) x (2000 kg)

8. The motion of a block sliding across a horizontal, frictionless surface can be described as one with
     a. a decreasing speed.
     b. an increasing speed.
   * c. a constant speed.
     d. a constant, non-zero acceleration.

9. If the force of friction on a child's wagon is 25 N, how much force must be applied to maintain a constant, non-zero velocity?
     a. 26 N
   * b. 25 N
     c. 24 N
     d. zero

10. There are three forces acting on an object: 6 N to the left, 5 N to the right, and 3 N to the left. What is the net force acting on the object?
     a. 4 N
   * b. 4 N left
     c. 4 N right
     d. 6 N left

11. What are the size and direction of the force that is the sum of a force of 3 N acting south and a force of 6 N acting north?
     a. 2 N north
   * b. 3 N north
     c. 6 N north
     d. 9 N north

12. What is the net force acting on an object which is under the influence of a 5 N force acting north and a 5 N force acting west?
     a. 5 N northwest
   * b. 7 N northwest
     c. 10 N northwest
     d. zero

13. What is the magnitude of the net force acting on an object which is under the influence of a 3 N force acting south and a 4 N force acting east?
    a. 3 N
    b. 4 N
  * c. 5 N
    d. 7 N

14. A subway train is moving with constant velocity along a level section of track. The net force on the first subway car is _____ the net force on the last subway car.
  * a. equal to
    b. much greater than
    c. slightly greater than
    d. less than

15. Forces of 4 N and 6 N act on an object. What is the minimum value for the sum of these two forces?
    a. zero
  * b. 2 N
    c. 4 N
    d. 10 N

16. You are analyzing a problem in which two forces act on an object. A 200-N force pulls to the right and a 40-N force pulls to the left. The net force acting on the object is
    a. 40 N to the left.
  * b. 160 N to the right.
    c. 200 N to the right.
    d. 240 N to the right.

17. If the net force on an object is directed due west, which way does the acceleration point?
  * a. due west
    b. due east
    c. west only if the velocity is west
    d. It could be in any westerly direction.

18. If the net force on a hot-air balloon is directed straight upward, which way does the acceleration point?
    a. downward
  * b. upward
    c. upward only if the balloon is rising
    d. upward and downwind

19. What kind of motion does a constant, non-zero net force produce on an object of constant mass?
    - a. constant speed
    - * b. constant acceleration
    - c. increasing acceleration
    - d. decreasing acceleration

20. If a body is acted on by a force of 10 N and doesn't accelerate, we have to assume
    - a. nothing. That's what should happen.
    - b. its inertia is too large.
    - * c. that the net force acting on the body is zero.
    - d. that the law of inertia only holds for large forces.

21. If you push on a railroad boxcar with a force of 300 N and it doesn't move, you can conclude that
    - a. Newton's second law is not valid.
    - b. this force is canceled by the third law force.
    - c. the boxcar has too much mass to accelerate.
    - * d. there is a total force of 300 N in the opposite direction.

22. You are applying a 400-N force to a freezer full of chocolate chip ice cream in an attempt to move it across the basement. It will not budge. The weight of the freezer (including the ice cream) is 1000 N. The friction force exerted by the floor on the freezer is
    - * a. 400 N.
    - b. greater than 400 N but less than 1000 N.
    - c. 1000 N.
    - d. greater than 1000 N.

23. When the same net force is applied to two blocks, the yellow one has a larger acceleration than the blue one. Which of the following is correct?
    - a. The yellow block has a larger mass.
    - * b. The blue block has a larger mass.
    - c. They have the same mass.

24. The same net force is applied to two blocks. If the yellow one has a larger mass than the blue one, which one will have the larger acceleration?
    - a. the yellow block
    - * b. the blue block
    - c. They have the same acceleration.

Chapter 3   Explaining Motion

25. What net force is required to accelerate 20 kg at 5 m/s^2.
    * a. 100 N
       b. 25 N
       c. 15 N
       d. 4 N

26. What acceleration is produced by a force of 30 N acting on a mass of 10 kg?
    * a. 3 m/s^2
       b. 10 m/s^2
       c. 30 m/s^2
       d. 300 m/s^2

27. What net force is needed to accelerate a 60-kg ice skater at 2 m/s^2?
       a. zero
       b. 30 N
       c. 60 N
    * d. 120 N

28. What is the mass of a cart that has an acceleration of 4 m/s^2 when a net force of 2000 N is applied to it?
       a. 8,000 kg
       b. 2000 kg
       c. 5000 kg
    * d. 500 kg

29. What acceleration is produced by a force of 100 N acting on a mass of 10 kg if its velocity is 20 m/s and the frictional force is 30 N?
       a. 10 m/s^2
       b. 9 m/s^2
       c. 8 m/s^2
    * d. 7 m/s^2

30. When the same net force is applied to object A and object B, object A has an acceleration three times that of object B. Which of the following is correct?
       a. Object A has three times the mass of object B.
    * b. Object A has one-third the mass of object B.
       c. Object A has a different, less streamlined shape than object B.
       d. Object A has more friction than object B.

31. If the mass and weight of an astronaut are measured on Earth and on the Moon, we will find that the masses are _____ and the weights are _____.
    a. the same ... the same
    b. different ... different
    * c. the same ... different
    d. different ... the same

32. The strength of gravity on the Moon is only 1/6th that on Earth. If an astronaut has a mass of 90 kg on earth, what would her mass be on the Moon?
    a. 540 kg
    * b. 90 kg
    c. 15 kg
    d. 6 kg

33. Which of the following is not a vector quantity?
    a. force
    b. acceleration
    c. weight
    * d. mass

34. The strength of gravity on the Moon is only 1/6th that on Earth. If an astronaut has a mass of 90 kg on Earth, what would her weight be on the Moon?
    a. 900 N
    * b. 150 N
    c. 90 N
    d. 15 N

35. The strength of gravity on Mars is only 40% of that on Earth. If a child has a mass of 40 kg on Earth, what would the child's weight be on Mars?
    a. 16 N
    b. 40 N
    * c. 160 N
    d. 400 N

36. An astronaut on a strange planet has a mass of 80 kg and a weight of 240 N. What is the value of the acceleration due to gravity on this planet?
    * a. 3 m/s^2
    b. 8 m/s^2
    c. 10 m/s^2
    d. 1/3 m/s^2

Chapter 3   Explaining Motion

37. A ball with a weight of 20 N is thrown vertically upward. What are the size and direction of the force on the ball just as it reaches the top of its path?
       a. zero
       b. 10 N upward
       c. 10 N downward
   *   d. 20 N downward

38. A ball with a weight of 20 N is thrown vertically upward. What is the acceleration of the ball just as it reaches the top of its path?
       a. zero
   *   b. 10 m/s^2 downward
       c. 10 m/s^2 upward
       d. The acceleration cannot be determined.

39. A ball falling from a great height will reach terminal speed when its _____ goes to zero.
       a. inertia
   *   b. net force
       c. weight
       d. speed

40. When a snowflake falls, it quickly reaches a terminal velocity. This happens because
       a. the mass of the snowflake is too small for gravity to have any effect.
   *   b. there is no net force acting on the snowflake.
       c. the snowflake has no weight.
       d. the mass of the snowflake is smaller than its weight.

41. A parachutist reaches terminal speed when
       a. her weight goes to zero.
   *   b. the force of air resistance equals her weight.
       c. the force of air resistance exceeds her weight.
       d. the force of air resistance equals her mass.

42. Two steel balls have the same size and shape, but one is hollow. They are dropped in air and their terminal speeds are measured. Which of the following statements is correct?
   *   a. The hollow ball has a smaller terminal speed because it requires a smaller air resistance to cancel the gravitational force on it.
       b. The hollow ball has a larger terminal speed because it requires a smaller air resistance to cancel the gravitational force on it.
       c. The terminal speeds are the same because the acceleration of gravity doesn't depend on mass.
       d. The terminal speeds are the same and equal to 10 m/s.

26

43. A professor decides to simulate the effects of air resistance by simultaneously dropping two balls in a long column of water. The two balls have the same size but have masses of 1 kg and 2 kg. What happens?
    a. The 1-kg ball hits the bottom first.
 * b. The 2-kg ball hits the bottom first.
    c. They hit the bottom at the same time.

44. A 40-kg crate is being pushed across a horizontal floor. If the coefficient of sliding friction is 0.3, what is the frictional force acting on the crate?
    a. 12 N
    b. 40 N
 * c. 120 N
    d. 400 N

45. A crate has a mass of 24 kg. What applied force is required to produce an acceleration of 3 m/s^2 if the frictional force is known to be 90 N?
    a. 72 N
    b. 90 N
 * c. 162 N
    d. 240 N

46. A 40-kg crate is being pushed across a horizontal floor by a horizontal force of 240 N. If the coefficient of sliding friction is 0.5, what is the acceleration of the crate?
    a. zero
 * b. 1 m/s^2
    c. 3 m/s^2
    d. 6 m/s^2

47. You apply a 75-N force to pull a child's wagon across the floor at a constant speed of 0.5 m/s. If you increase your pull to 80 N, the wagon will
    a. continue to move at 0.5 m/s.
    b. speed up briefly and then move at a faster constant speed.
 * c. continue to speed up.

48. If F_1 is the force exerted on a cart by a horse and F_2 is the force exerted on the horse by the cart, then F_1 is _____ F_2.
    a. much greater than
    b. slightly greater than
 * c. equal to
    d. slightly less than

Chapter 3   Explaining Motion

49. If Earth exerts a gravitational force of 20,000 N on a satellite in synchronous orbit, what force does the satellite exert on Earth?
   - a. zero
   - b. a small fraction of 1 N
   - c. 5000 N
   - * d. 20,000 N

50. A ball with a weight of 40 N is falling freely toward the surface of the Moon. What force does this ball exert on the Moon?
   - a. zero
   - b. 40 N down
   - * c. 40 N up

51. Which of the following is the third-law force that accompanies the force that an apple exerts on a tree? It is the force that
   - a. Earth exerts on the apple.
   - b. the apple exerts on Earth.
   - * c. the tree exerts on the apple.
   - d. the air exerts on the apple.

52. A book sits at rest on a table. Which force does Newton's third law tell us is equal and opposite to the gravitational force acting on the book?
   - a. the normal force by the table on the book
   - b. the normal force by the book on the table
   - * c. the gravitational force by the book on Earth
   - d. the net force on the book

53. You leap from a bridge with a bungee cord tied around your ankles. As you approach the river below, the bungee cord begins to stretch and you begin to slow down. The force of the cord on your ankles to slow you is _____ the force of your ankles on the cord to stretch it?
   - a. less than
   - * b. equal to
   - c. greater than

54. You leap from a bridge with a bungee cord tied around your ankles. As you approach the river below, the bungee cord begins to stretch and you begin to slow down. The force of the cord on your ankles to slow you is _____ your weight?
   - a. less than
   - b. equal to
   - * c. greater than

55. Because the forces demanded by Newton's third law of motion are equal in magnitude and opposite in direction, how can anything ever be accelerated?
      a. Newton's third law only applies when there is NO acceleration.
*   b. The forces in question act on different bodies.
      c. Newton's second law is more important.
      d. Third law forces cannot cause accelerations.

56. Two skaters face each other on perfectly smooth ice. One skater has twice the mass as the other. Assuming that the bigger skater pushes the smaller, which of the following statements is true?
      a. The bigger skater won't move.
*   b. The bigger skater will move with the smaller acceleration.
      c. The bigger skater will move with the larger acceleration.
      d. Both skaters move with the same acceleration.

57. Terry and Chris pull hand-over-hand on opposite ends of a rope while standing on a frictionless frozen pond. Terry's mass is 75 kg and Chris's mass is 50 kg. If Terry's acceleration is 2 m/s^2, what is Chris's acceleration?
      a. 2 m/s^2
*   b. 3 m/s^2
      c. 6 m/s^2
      d. 10 m/s^2

58. A child stands on a bathroom scale while riding in an elevator. The child's weight when the elevator is not moving is 500 N. What does the scale read when the elevator accelerates upward while traveling downward?
*   a. greater than 500 N
      b. less than 500 N
      c. equal to 500 N

59. A child stands on a bathroom scale while riding in an elevator. The child's weight when the elevator is not moving is 500 N. What does the scale read when the elevator free-falls from the tenth floor?
      a. greater than 500 N
*   b. less than 500 N
      c. equal to 500 N

60. A child stands on a bathroom scale while riding in an elevator. The child's weight when the elevator is not moving is 500 N. What does the scale read when the elevator is moving at a constant speed?
    a. greater than 500 N
    b. less than 500 N
    *   c. equal to 500 N

61. A child stands on a bathroom scale while riding in an elevator. The child's weight when the elevator is not moving is 500 N. What does the scale read as the elevator comes to a stop at the tenth floor on its upward trip?
    a. greater than 500 N
    *   b. less than 500 N
    c. equal to 500 N

62. You are riding an elevator from your tenth-floor apartment to the parking garage in the basement. As you approach the garage, the elevator begins to slow. The net force acting on you is
    a. always equal to your weight
    *   b. directed upward
    c. directed downward
    d. zero

63. You are riding an elevator from the parking garage in the basement to the tenth floor of an apartment building. As you approach your floor, the elevator begins to slow. The net force acting on you is
    a. always equal to your weight
    b. directed upward
    *   c. directed downward
    d. zero

64. If you stand on a spring scale in your bathroom at home, it reads 600 N, which means your mass is 60 kg. If instead you stand on the scale while accelerating at 2 m/s^2 upward in an elevator, what would the scale read?
    a. 120 N
    b. 480 N
    c. 600 N
    *   d. 720 N

Chapter 4: Motions in Space

1. If a race car is traveling around a circular track at a constant speed of 100 mph, we know that the car experiences
   - a. no net force.
   * b. a centripetal force.
   - c. a centrifugal force.
   - d. a net force in the forward direction.

Questions 2, 3, 4, and 5: A racecar is moving counterclockwise on a circular path as shown in the diagram. Imagine that at this instant, the car is at point P and moving at a speed of 100 mph.

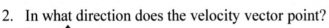

2. In what direction does the velocity vector point?
   * a. ↑
   - b. ↓
   - c. →
   - d. ←

3. In what direction does the net force point?
   - a. ↑
   - b. ↓
   - c. →
   * d. ←

4. In what direction does the acceleration point?
   - a. ↑
   - b. ↓
   - c. →
   * d. ←

5. Imagine that the car hits a large oil slick that reduces the force of friction to zero. In what direction will the car slide?
   * a. ↑
   - b. ↓
   - c. →
   - d. ←

Chapter 4   Motions in Space

6. A bug rides on a phonograph record. In which direction does the acceleration of the bug point?

        a. tangent to the circular path

    *   b. toward the center of the record

        c. away from the center of the record

        d. up

7. A bug rides on a phonograph record. In which direction does the change in velocity of the bug point?

        a. tangent to the circular path

    *   b. toward the center of the record

        c. away from the center of the record

        d. up

8. Your instructor rides on a merry-go-round turning at a constant rate. In which direction does the net force on your instructor point?

        a. tangent to the circular path

    *   b. toward the center

        c. away from the center

        d. down

9. In straight line motion the

    *   a. acceleration is parallel (or antiparallel) to the velocity.

        b. acceleration is perpendicular to the velocity.

        c. acceleration is vertical, while the velocity can be in any direction.

        d. acceleration is vertical and the velocity is horizontal.

10. In uniform circular motion the

        a. acceleration is parallel (or antiparallel. to the velocity.

    *   b. acceleration is perpendicular to the velocity.

        c. acceleration is vertical, while the velocity can be in any direction.

        d. acceleration is vertical and the velocity is horizontal.

11. What is the change in velocity if a car traveling 30 m/s north slows to 20 m/s north?

        a. 20 m/s north

        b. 10 m/s north

    *   c. 10 m/s south

        d. 30 m/s south

12. What is the change in velocity if a car traveling 30 m/s north changes to 20 m/s south?
    a. 10 m/s north
    b. 20 m/s south
    c. 30 m/s north
    * d. 50 m/s south

13. An airplane is flying south at 40 m/s. What is the magnitude of the airplane's change in velocity if it is later flying west at 30 m/s?
    a. 10 m/s
    b. 30 m/s
    c. 40 m/s
    * d. 50 m/s

14. A car is traveling south at 30 m/s. Later it is observed traveling west at 30 m/s. What is the car's change in velocity?
    a. 42 m/s north
    b. 42 m/s west
    c. 42 m/s southwest
    * d. 42 m/s northwest

15. A migrating bird is initially flying south at 8 m/s. To avoid hitting a high-rise building, the bird veers and changes its velocity to 6 m/s east over a period of 2 s. What is the magnitude of the bird's average acceleration during this 2-s interval?
    * a. 5 m/s^2
    b. 7 m/s^2
    c. 10 m/s^2
    d. 14 m/s^2

16. A bunny is initially hopping east at 8 m/s when it first sees a fox. Over the next half second, the bunny changes its velocity to west at 12 m/s and escapes. What was the magnitude of the bunny's average acceleration during this half-second interval?
    a. 4 m/s^2
    b. 8 m/s^2
    c. 20 m/s^2
    * d. 40 m/s^2

17. By what factor does the centripetal acceleration change if a car goes around a corner twice as fast?
    a. 0.5
    b. It stays the same.
    c. 2
    * d. 4

Chapter 4   Motions in Space

18. What centripetal acceleration is required to follow a circular path with a radius of 50 m at a speed of 20 m/s?
    a. 2 m/s^2
    b. 4 m/s^2
    c. 6 m/s^2
    * d. 8 m/s^2

19. A 60-kg person on a merry-go-round is traveling in a circle with a radius of 3 m at a speed of 3 m/s. What is the magnitude of the acceleration experienced by this person?
    a. 1 m/s^2
    * b. 3 m/s^2
    c. 10 m/s^2
    d. 180 m/s^2

20. A 60-kg person on a merry-go-round is traveling in a circle with a radius of 3 m at a speed of 3 m/s. What is the magnitude of the net force experienced by this person?
    a. zero
    b. 60 N
    * c. 180 N
    d. 540 N

21. A cyclist turns a corner with a radius of 50 m at a speed of 20 m/s. What is the magnitude of the cyclist's acceleration?
    a. 0.4 m/s^2
    b. 2.5 m/s^2
    * c. 8 m/s^2
    d. 10 m/s^2

Questions 22, 23, 24, and 25: A gun is held horizontally and fired. At the same time the bullet leaves the gun's barrel an identical bullet is dropped from the same height. Neglect air resistance.

22. Which bullet will hit the ground first?
    a. The bullet that was fired.
    b. The bullet that was dropped.
    * c. It will be a tie.

23. Which bullet will hit the ground with the greater velocity?
    * a. The bullet that was fired.
    b. The bullet that was dropped.
    c. It will be a tie.

24. Which would hit the ground first if this experiment were conducted on the surface of the Moon?
    a. The bullet that was fired.
    b. The bullet that was dropped.
    * c. It will be a tie.

25. If the bullets were not identical, but rather the dropped bullet had twice the mass of the other, which bullet would hit the ground first?
    a. The bullet that was fired.
    b. The bullet that was dropped.
    * c. It will be a tie.

26. A 2-kg ball is thrown horizontally at a speed of 10 m/s. At the same time a 1-kg ball is dropped from the same height. Ignoring air resistance, which ball hits the ground first?
    a. the 1-kg ball
    b. the 2-kg ball
    * c. It's a tie.

27. A red ball is thrown straight down from the edge of a tall cliff with a speed of 30 m/s. At the same time a green ball is thrown straight up with the same speed. How many seconds later than the red ball will the green ball land?
    a. 3 s
    * b. 6 s
    c. 10 s
    d. There is not enough information to say.

28. A red ball is thrown straight down from the edge of a tall cliff with a speed of 30 m/s. At the same time a green ball is thrown straight up with the same speed. Which ball (if either) will be traveling faster when it reaches the ground below?
    a. the red ball
    b. the green ball
    * c. Both balls will have the same speed.

29. Which of the following statements about projectile motion is true?
    a. The horizontal and vertical motions are independent.
    b. The force on the projectile is constant throughout the flight.
    c. The acceleration of the projectile is constant throughout the flight.
    * d. All of the above statements are true.

Chapter 4   Motions in Space

30. In projectile motion the
      a. acceleration is parallel (or antiparallel) to the velocity.
      b. acceleration is perpendicular to the velocity.
  * c. acceleration is vertical, while the velocity can be in any direction.
      d. acceleration is vertical and the velocity is horizontal.

31. A baseball player throws a ball from left field toward home plate. Assume that you can neglect the effects of air resistance. At the instant the ball reaches its highest point, what is the direction of the ball's velocity?
      a. up
      b. down
  * c. horizontal
      d. The velocity is zero.

32. A football quarterback throws a long pass toward the end zone. Assume that you can neglect the effects of air resistance. At the instant the ball reaches its highest point, what is the direction of the net force on the ball?
      a. up
  * b. down
      c. horizontal
      d. The net force is zero.

33. A baseball player throws a ball from left field toward home plate. Assume that you can neglect the effects of air resistance. At the instant the ball reaches its highest point, what is the direction of the ball's acceleration?
      a. up
  * b. down
      c. horizontal
      d. The acceleration is zero.

34. A football quarterback throws a long pass toward the end zone. Assume that you can neglect the effects of air resistance. At the instant the ball reaches its highest point, what is the acceleration of the ball?
      a. zero
  * b. 10 m/s^2 downward
      c. 10 m/s^2 upward

35. A physics student reports that upon arrival on planet X, he promptly sets up the "monkey-shoot" demonstration. If the gravity on planet X is twice what it is on Earth, he should obtain a
    a. miss because the monkey's weight is twice as big on planet X.
    b. hit only if the ball's horizontal velocity is increased.
    c. miss because the monkey's mass is unchanged.
    * d. hit because the ball and the monkey fall vertically with the same acceleration.

36. A baseball is hit with a horizontal speed of 22 m/s and a vertical speed of 14 m/s upward. What are these speeds 1 s later?
    * a. 22 m/s horizontal and 4 m/s upward
    b. 22 m/s horizontal and 24 m/s upward
    c. 12 m/s horizontal and 4 m/s upward
    d. 12 m/s horizontal and 14 m/s upward

37. After being hit, a baseball has a horizontal speed of 20 m/s and a vertical speed of 25 m/s upward. Ignoring air resistance what are these speeds 1 s later?
    a. 20 m/s horizontal and 25 m/s vertical
    * b. 20 m/s horizontal and 15 m/s vertical
    c. 10 m/s horizontal and 25 m/s vertical
    d. 20 m/s horizontal and 35 m/s vertical

38. A rock is thrown off a tall cliff with a vertical speed of 25 m/s upward and a horizontal speed of 30 m/s. What will these speeds be 3 s later?
    a. 25 m/s upward and 30 m/s horizontal
    * b. 5 m/s downward and 30 m/s horizontal
    c. 25 m/s upward and 0 m/s horizontal
    d. 30 m/s downward and 60 m/s horizontal

39. An ashtray slides across a table with a speed of 0.9 m/s and falls off the edge. If it takes 0.4 s to reach the floor, how far from the edge of the table does the ashtray land?
    * a. 0.36 m
    b. 0.4 m
    c. 0.9 m
    d. 1.3 m

40. A car drives off a vertical cliff at a speed of 24 m/s. If it takes 3 s for the car to hit the ground, how far from the base of the cliff does it land?
    a. 3 m
    b. 8 m
    c. 24 m
    * d. 72 m

Chapter 4   Motions in Space

41. A rock is thrown off a tall cliff with a vertical speed of 25 m/s upward and a horizontal speed of 30 m/s. If the rock lands 8 s later, how far from the base of the cliff will it land?
    a. 30 m
    b. 120 m
    c. 165 m
    * d. 240 m

42. A bowling ball rolls off the edge of a giant's table at 15 m/s. If it takes 4 s for the ball to hit the ground, how far does it land from the base of the table?
    a. 10 m
    b. 15 m
    c. 40 m
    * d. 60 m

43. A bowling ball rolls off the edge of a giant's table at 15 m/s. If it takes 4 s for the ball to hit the ground, what is the height of the table?
    a. 40 m
    b. 60 m
    * c. 80 m
    d. 160 m

44. Angel Falls in southeastern Venezuela is the highest uninterrupted waterfall in the world, dropping 979 m (3212 ft). Ignoring air resistance, it would take 14 s for the water to fall from the lip of the falls to the river below. If the water lands 50 m from the base of the vertical cliff, what was its horizontal speed at the top?
    * a. 3.6 m/s
    b. 9.8 m/s
    c. 50 m/s
    d. 700 m/s

45. A baseball is hit with a vertical speed of 10 m/s and a horizontal speed of 30 m/s. How long will the ball remain in the air?
    a. 1 s
    * b. 2 s
    c. 3 s
    d. 6 s

46. A tennis ball is hit with a vertical speed of 10 m/s and a horizontal speed of 30 m/s. How far will the ball travel horizontally before landing?
    a. 30 m
    b. 40 m
    * c. 60 m
    d. 80 m

Chapter 5: Gravity

1. What force drives the planets along their orbits?
   - a. gravity
   - b. magnetism
   - c. solar wind
   - * d. No force is needed to drive them along their orbits.

2. Assuming that a planet is in a circular orbit about the Sun, what is the force that drives the planet along its orbit; that is, tangent to the circle?
   - a. gravity
   - b. magnetism
   - c. solar wind
   - * d. No force is needed to drive it along its orbit.

3. The discussion of the "launched apple" and the Moon shows that the
   - a. velocities of the apple and Moon are constant.
   - * b. motion of the Moon and apple can be explained by the same laws.
   - c. apple and the Moon have the same acceleration.
   - d. apple and the Moon experience the same-sized force.

4. If we imagine launching an apple into a circular orbit about Earth and ignore the effects of air resistance, we know that the apple will experience
   - a. a constant velocity.
   - b. no net force.
   - c. a force due to its inertia.
   - * d. a centripetal force due to gravity.

5. Which of the following statements about the Moon is correct?
   - a. The Moon has a constant velocity.
   - b. There is no net force acting on the Moon.
   - c. Earth exerts a stronger force on the Moon than the Moon exerts on Earth.
   - * d. The Moon experiences a centripetal acceleration toward Earth.

6. Which of the following statements about Venus is correct?
   - a. Venus has a constant velocity.
   - b. There is no net force acting on Venus.
   - c. The Sun exerts a stronger force on Venus than Venus exerts on the Sun.
   - * d. Venus is continually accelerating toward the Sun.

Chapter 5   Gravity

7. Which of the following statements about the Moon is <u>not</u> correct?
    a. The acceleration due to gravity on the Moon is weaker than on Earth.
    b. Earth's gravitational pull on the Moon equals the Moon's gravitational pull on Earth.
    c. There is a net force acting on the Moon.
    * d. The Moon is not accelerating.

8. Which of the following statements about Venus is <u>not</u> correct?
    a. The Sun's gravitational pull on Venus equals Venus' gravitational pull on the Sun.
    b. There is a net force acting on Venus.
    c. Venus is accelerating toward the Sun.
    * d. There is no gravity on the surface of Venus.

9. The size of the gravitational force that Earth exerts on the Moon is _____ that the Moon exerts on Earth.
    a. greater than
    * b. the same as
    c. smaller than

10. Earth is held in its orbit by the gravitational force of the Sun. Therefore, the force that the Sun exerts on Earth is _____ that Earth exerts on the Sun.
    a. greater than
    b. smaller than
    * c. the same as

11. The gravitational attraction of the Sun for Earth is _____ that of Earth for the Sun.
    * a. the same as
    b. greater than
    c. smaller than

12. What is the acceleration due to Earth's gravity at a distance of one Earth radius above Earth's surface?
    * a. 2.5 m/s^2
    b. 5 m/s^2
    c. 10 m/s^2
    d. 20 m/s^2

13. What is the acceleration due to Earth's gravity at a distance of 10 Earth radii from Earth's center?
    a. 10 m/s^2
    b. 1 m/s^2
    * c. 0.1 m/s^2
    d. 0.01 m/s^2

14. An astronaut weighs 900 N when measured on the surface of Earth. How large would the force of gravity on him be if he were in an Earth satellite at an altitude equal to Earth's radius?
*   a. 225 N
    b. 450 N
    c. 900 N
    d. 3600 N

15. Al the astronaut has a weight of 800 N when he is standing on the surface of Earth. What is the force of gravity acting on him when he is in a space station orbiting Earth at a distance of three Earth radii above the surface?
    a. 800 N
    b. 200 N
    c. 100 N
*   d. 50 N

16. If you double the length of each side of a cube, its volume increases by what factor?
    a. 2
    b. 4
    c. 6
*   d. 8

17. If you double the length of each side of a cube, its surface area increases by what factor?
    a. 2
*   b. 4
    c. 6
    d. 8

18. If you double the radius of a sphere, its volume increases by what factor?
    a. 2
    b. 4
    c. 6
*   d. 8

19. If you double the radius of a sphere, its surface area increases by what factor?
    a. 2
*   b. 4
    c. 6
    d. 8

Chapter 5   Gravity

20. A future space traveler, Skip Parsec, lands on the planet MSU3, which has the same mass as Earth but twice the radius. If Skip weighs 800 N on Earth's surface, how much does he weigh on MSU3's surface?
    a. 50 N
    b. 100 N
    *   c. 200 N
    d. 400 N

21. Astronaut Skip Parsec visits planet MSU8, which is composed of the same materials as Earth, but has twice the radius. If Skip weighs 800 N on Earth's surface, how much does he weigh on MSU8's surface?
    a. 400 N
    b. 800 N
    *   c. 1600 N
    d. 3200 N

22. In the law of universal gravitation the force _____ as the mass increases and _____ as the distance increases.
    a. increases ... increases
    b. decreases ... increases
    *   c. increases ... decreases
    d. decreases ... decreases

23. The gravitational force between two metal spheres in outer space is 2000 N. How large would the force be if the two spheres were four times as far apart?
    a. 32,000 N
    b. 8000 N
    c. 500 N
    *   d. 125 N

24. The gravitational force between two metal spheres in outer space is 1800 N. How large would the force be if the two spheres were twice as far apart?
    a. 7200 N
    b. 3600 N
    c. 900 N
    *   d. 450 N

25. The gravitational force between two metal spheres in outer space is 1000 N. How large would this force be if each of the two spheres had twice the mass?
    a. 1000 N
    b. 2000 N
*   c. 4000 N
    d. 16,000 N

26. A solid lead sphere of radius 10 m (about 66 ft across!) has a mass of about 57 million kg. If two of these spheres are floating in deep space with their centers 20 m apart, the gravitational attraction between the spheres is only 540 N (about 120 lb). How large would this gravitational force be if the distance between the centers of the two spheres were tripled?
*   a. 60 N
    b. 180 N
    c. 1620 N
    d. 4860 N

27. Two spacecraft in outer space attract each other with a force of 20 N. What would the attractive force be if they were one-half as far apart?
    a. 5 N
    b. 10 N
    c. 40 N
*   d. 80 N

28. Does the Moon orbit the Sun?
    a. Yes. It goes in a circle about the Sun.
    b. No. It orbits Earth.
*   c. Yes, but it also orbits Earth.
    d. No, but it would if Earth were not present.

29. During the Apollo flights to the Moon a well-known TV newscaster made the following statement, "The Apollo space craft is now leaving the gravitational force of Earth." This statement is incorrect. He should have said that the space craft
    a. was attracted only by the Moon.
    b. was attracted only by the Sun
*   c. was attracted more by the Moon than by Earth.
    d. entered a region of space where there were no gravitational forces.

30. The numerical value of G, the gravitational constant, was determined
    a. from knowledge of Earth's mass.
    b. from the law of universal gravitation and the value of the acceleration due to gravity.
    c. from the value of the Moon's acceleration.
*   d. by measuring the force between masses in the laboratory.

Chapter 5   Gravity

31. The gravitational constant G in the law of universal gravitation
    a. is believed to be constant with time.
    b. is believed to have the same value throughout space.
    c. leads to a determination of the mass of Earth.
  * d. All of the above are true.

32. When Cavendish claimed that he "weighed" Earth, he actually calculated the
    a. force that the Moon exerts on Earth.
    b. weight of Earth.
  * c. mass of Earth.
    d. force that the Sun exerts on Earth.

33. The law of universal gravitation is written $F = GMm/r^2$. Why did we use the form $F = mg$ when we studied projectile motion?
    a. The first form is not valid for projectile motion.
    b. The first form does not work because it requires two masses.
    c. The first form is not valid near the surface of Earth.
  * d. The second form is a good approximation to the first and much simpler to use.

34. Which of the following would NOT cause the gravitational force on an object near Earth's surface to increase?
    a. an ore deposit just under the surface
    b. a lower elevation
    c. an increase in its mass
  * d. a horizontal velocity

35. Earth exerts a gravitational force of 7000 N on one of the communications satellites. What force does the satellite exert on Earth?
    a. more than 7000 N
  * b. 7000 N
    c. less than 7000 N
    d. zero

36. An astronaut weighs 800 N when measured on Earth's surface. How large would his weight be if he were in a satellite at an altitude equal to Earth's radius?
  * a. zero
    b. less than 800 N
    c. 800 N
    d. more than 800 N

37. If an astronaut with a weight of 800 N on Earth steps on a bathroom scale while he is in Earth orbit, the scale will read
  * a. zero.
    b. less than 800 N.
    c. 800 N.
    d. more than 800 N.

38. Al the astronaut has a mass of 90 kg and a weight of 900 N when he is standing on Earth's surface. What is his mass when he is in a space station orbiting Earth with a radius of three Earth radii?
    a. zero
  * b. 90 kg
    c. 30 kg
    d. 10 kg

39. From film taken in SkyLab and the Space Shuttle, we learned that objects in SkyLab
  * a. have mass but no weight.
    b. have mass but no force due to gravity.
    c. have neither mass nor weight.
    d. fall to the floor with an acceleration of 9.5 m/s^2.

40. Over which of the following locations is it possible to have a synchronous satellite?
    a. New York City
    b. Los Angeles
    c. North Pole
  * d. equator

41. Communications satellites are synchronous satellites that orbit Earth each
    a. 90 minutes
  * b. 24 hours
    c. 28 days
    d. They don't orbit Earth; they just stay in one place.

Questions 42, 43, and 44: Sally is an astronaut who has a mass of 60 kg. Currently she is conducting experiments in a permanent space station that is orbiting Earth at an altitude equal to Earth's radius.

42. What is Sally's mass as measured in the space station?
    a. zero
    b. 15 kg
    c. 30 kg
  * d. 60 kg

Chapter 5   Gravity

43. What is the force of gravity acting on Sally while she is in the space station?
    a. zero
  * b. 150 N
    c. 300 N
    d. 600 N

44. What is Sally's weight in the space station?
  * a. zero
    b. 150 N
    c. 300 N
    d. 600 N

45. The highest high tides and the lowest low tides occur when the Moon is
  * a. full or new.
    b. full only.
    c. new only.
    d. first and third quarter.

46. In a typical 24 hour day, there are
    a. one high tide and one low tide.
    b. one high tide and two low tides.
    c. two high tides and one low tide.
  * d. two high tides and two low tides.

47. Which of the following celestial bodies has the greatest influence on Earth's tides?
  * a. Moon
    b. Sun
    c. Venus
    d. Jupiter

48. Although Earth's gravitational influence gets weaker with distance, we believe that it extends as far as the
    a. Sun.
    b. edge of the solar system.
    c. edge of the galaxy.
  * d. universe extends.

49. According to the law of universal gravitation the Earth's gravitational field extends
    a. to the Sun.
    b. to the edge of the solar system.
    c. to the edge of the galaxy.
  * d. throughout the universe.

50. What is the magnitude of Earth's gravitational field at a distance equal to twice Earth's radius?
    a. 20 N/kg
    b. 10 N/kg
    c. 5 N/kg
    * d. 2.5 N/kg

51. Given that the acceleration due to gravity on Mars is 3.7 m/s^2, what is the gravitational field near Mar's surface?
    a. 3.7 N/kg
    b. 3.7 N/kg up
    * c. 3.7 N/kg down
    d. 10 N/kg down

52. If Earth's mass were suddenly and magically reduced to half its present value, the Sun's gravitational force on Earth would
    a. be reduced by a factor of 4.
    * b. be reduced by a factor of 2.
    c. remain the same.
    d. increase by a factor of 2.

53. If Earth's mass were suddenly and magically reduced to half its present value, the magnitude of Earth's acceleration about the Sun would
    a. be reduced by a factor of 4.
    b. be reduced by a factor of 2.
    * c. remain the same.
    d. increase by a factor of 2.

54. The gravitational force between two very large metal spheres in outer space is 200 N. How large would this force be if the mass of each sphere were cut in half?
    a. 12.5 N
    * b. 50 N
    c. 100 N
    d. 200 N

55. A 320-kg satellite experiences a gravitational force of 800 N. What is the radius of the satellite's orbit? (Earth's radius is 6,400 km.)
    a. 6,400 km
    * b. 12,800 km
    c. 19,200 km
    d. 25,600 km

56. A 320-kg satellite experiences a gravitational force of 800 N. What is the satellite's altitude? (Earth's radius is 6,400 km.)
    * a. 6,400 km
      b. 12,800 km
      c. 19,200 km
      d. 25,600 km

57. A 600-kg geosynchronous satellite has an orbital radius of 6.6 Earth radii. What gravitational force does Earth exert on the satellite?
      a. 13.8 N
      b. 90.9 N
    * c. 138 N
      d. 909 N

58. What is the gravitational force between two 20-kg iron balls separated by a distance of 0.5 m? The gravitational constant is $G = 6.67 \times 10^{-11}$ N-m^2/kg^2.
      a. 2.67×10^{-9} N
      b. 5.34×10^{-9} N
      c. 5.34×10^{-8} N
    * d. 1.07×10^{-7} N

59. The masses of the Moon and Earth are 7.4×10^{22} kg and 6×10^{24} kg, respectively. Earth–Moon distance is 3.8×10^8 m. What is the size of the gravitational force between Earth and the Moon? The gravitational constant is $G = 6.67 \times 10^{-11}$ N-m^2/kg^2.
    * a. 2.05×10^{20} N
      b. 7.79×10^{28} N
      c. 3.07×10^{30} N
      d. 1.17×10^{39} N

60. If an astronaut in full gear has a weight of 1200 N on Earth, how much will the astronaut weigh on the Moon?
      a. 20 N
      b. 120 N
    * c. 200 N
      d. 720 N

61. The acceleration due to gravity on Titan, Saturn's largest Moon, is about 1.4 m/s^2. What would a 60-kg scientific instrument weigh on Titan?
      a. 43 N
      b. 60 N
    * c. 84 N
      d. 600 N

62. Mercury has a radius of about 0.38 Earth radii and a mass of only 0.055 Earth masses. Estimate the acceleration due to gravity on Mercury.
    a. 1.45 m/s^2
  * b. 3.81 m/s^2
    c. 26.3 m/s^2
    d. 69.1 m/s^2

63. An 90-kg satellite orbits a distant planet with a radius of 4000 km and a period of 280 min. From the radius and period, you calculate the satellite's acceleration to be 0.56 m/s^2. What is the gravitational force on the satellite?
  * a. 50.4 N
    b. 90 N
    c. 720 N
    d. 12,000 N

Chapter 6: Momentum

1. Under what conditions is mass conserved?
   - a. Only when the change is a chemical one.
   - b. Only when the change is a physical one.
   - * c. Whenever the system is isolated or closed.
   - d. Always.

2. Which has the greater momentum, a heavy truck at rest or a moving roller skate?
   - a. Cannot tell from the information given.
   - b. the heavy truck
   - * c. the roller skate
   - d. The momenta are equal.

3. Which has the greater momentum, an 18-wheeler parked at the curb or a Volkswagen rolling down a hill?
   - a. 18-wheeler
   - * b. Volkswagen
   - c. The momenta are equal.
   - d. Could be either.

4. Linear momentum is defined to be
   - a. mass times speed.
   - * b. mass times velocity.
   - c. mass times acceleration.
   - d. weight times velocity.

5. If a sports car with a mass of 1000 kg travels down the road with a speed of 30 m/s, its momentum is 30,000
   - a. kg/(m/s)
   - b. kg·s/m
   - * c. kg·m/s
   - d. kg·m^2/s^2

6. If a sports car with a mass of 1000 kg travels down the road with a speed of 40 m/s, its momentum is
   - a. 100 kg/m/s
   - b. 20,000 kg·m/s
   - * c. 40,000 kg·m/s
   - d. 400,000 kg·m^2/s^2

7. How fast would you have to throw a baseball ($m = 145$ g) to give it the same momentum as a 10-g bullet traveling at 800 m/s?
    a. 0.55 m/s
    b. 1.8 m/s
    * c. 55 m/s
    d. 800 m/s

8. How fast would an 80-kg person need to run to have the same momentum as an 18-wheeler ($m = 24{,}000$ kg) rolling along at 1 mph?
    * a. 300 mph
    b. 3000 mph
    c. 1,920 mph
    d. 1,920,000 mph

9. Newton's second law can be rearranged to show that the _____ is equal to the _____.
    a. momentum ... impulse
    b. change in momentum ... change in impulse
    * c. change in momentum ... impulse
    d. momentum ... change in impulse

10. Padded dashboards in automobiles are safer because the
    a. momentum change is less
    b. impulse is less
    c. impact time is less
    * d. impact time is greater

11. The stunt person who is shot by a bandit and falls from the balcony into an air bag rather than onto the ground will not be hurt because the
    a. momentum change is less for the air bag.
    b. momentum is less for the air bag.
    c. impulse is less for the air bag.
    * d. increased stopping time means a smaller stopping force.

12. Air bags are used by stunt people when they fall off buildings to reduce the _____ that occurs during the collision.
    a. change in momentum
    b. impulse
    * c. force
    d. change in velocity

Chapter 6   Momentum

13. The 12-ounce boxing gloves used in amateur fights hurt less than the 6-ounce gloves used in professional fights because the _____ stopping times mean _____ forces.
    * a. increased ... smaller
      b. decreased ... smaller
      c. increased ... larger
      d. decreased ... larger

14. Why is skiing into a wall of deep powder less hazardous to your health than skiing into a wall of bricks? Assume in both cases that you have the same initial speed and come to a complete stop.
      a. The change in momentum is less in powder.
      b. The impulse is less in powder.
    * c. The increased stopping time in powder means a smaller stopping force.
      d. The decreased stopping time in powder means a larger stopping force.

15. A tailgunner jumped from a Lancaster bomber but did not break any bones or die because he fell into the branches of a tree and then into a snow bank. Physics explains this because
      a. the change in momentum was less than hitting the ground directly.
      b. the impulse in less in trees and snow than ground.
    * c. the increased stopping time in the tree meant a smaller stopping force.
      d. the decreased stopping time in the tree meant a smaller stopping force.

16. An astronaut training at the Craters of the Moon in Idaho jumps off a platform in full spacewalk gear and hits the surface at 5 m/s. If later on the Moon the astronaut jumps from the LEM and hits the surface at the same speed, the impulse will be _____ that on Earth.
    * a. the same as
      b. larger than
      c. smaller than

17. Assuming that your author jumps off the roof of a garage and lands on the ground. How will the impulse the ground exerts on him if he lands on grass compare to that if he lands on concrete?
      a. The impulse will be larger if he lands on concrete.
      b. The impulse will be larger if he lands on grass.
    * c. The impulses will be the same independent of the surface.

18. Which of the following will cause the largest change in the momentum of an object? A force of _____ acting for _____.
      a. 3 N ... 6 s
      b. 4 N ... 5 s
    * c. 5 N ... 5 s
      d. 6 N ... 3 s

19. What is the impulse of a 5 N force acting for 15 s?
    a. 3 N·s
    b. 15 N·s
    c. 20 N·s
  * d. 75 N·s

20. What change in momentum occurs when a force of 20 N acts for 4 s?
    a. 5 kg·m/s
    b. 16 kg·m/s
    c. 24 kg·m/s
  * d. 80 kg·m/s

21. What impulse is need to stop a 1200-kg car traveling at 20 m/s?
    a. 60 N·s
    b. 240 N·s
    c. 1200 N·s
  * d. 24,000 N·s

22. A 1400-kg car has a speed of 20 m/s. What average force is required to stop the car in 10 s?
    a. 70 N
    b. 700 N
  * c. 2800 N
    d. 280,000 N

23. What average force is required to stop a 120-kg football player running at 8 m/s in a time of 0.4 s?
    a. 6 N
    b. 37.5 N
    c. 384 N
  * d. 2400 N

24. What average net force is needed to accelerate a 1200-kg car to a speed of 30 m/s in a time of 8 s?
    a. 320 N
  * b. 4500 N
    c. 144,000 N
    d. 288,000 N

25. It takes about 30 s for a jet plane to go from rest to the takeoff speed of 100 mph (44.7 m/s). What is the average horizontal force that the seat exerts on the back of a 80-kg passenger during takeoff?
    a. 16.8 N
    b. 53.7 N
 * c. 119 N
    d. 11,300 N

26. A coach is hitting pop flies to the outfielders. If the baseball (m = 145 g) stays in contact with the bat for 0.04 s and leaves the bat with a speed of 50 m/s, what is the average force acting on the ball?
 * a. 181 N
    b. 290 N
    c. 56,400 N
    d. 181,000 N

27. A tennis ball (m = 0.2 kg) is thrown at a brick wall. It is traveling horizontally at 16 m/s just before hitting the wall and rebounds from the wall at 8 m/s, still traveling horizontally. The ball is in contact with the wall for 0.04 s. What is the magnitude of the average force of the wall on the ball?
    a. 40 N
    b. 80 N
 * c. 120 N
    d. 640 N

28. A very hard rubber ball (m = 0.6 kg) is falling vertically at 6 m/s just before it bounces on the floor. The ball rebounds back at essentially the same speed. If the collision with the floor lasts 0.04 s, what is the average force exerted by the floor on the ball?
    a. 45 N
    b. 90 N
 * c. 180 N
    d. 360 N

29. If we examine a ball in free fall, we find that the momentum of the ball is not constant. This is not a violation of the law of conservation of momentum because
    a. the force of gravity acts on the ball.
    b. the ball experiences an external force.
    c. the system is not closed.
 * d. All of the above.

Physics: A World View, Sixth Edition by Larry Kirkpatrick and Gregory Francis

30. If rockets are fired from an airplane in the forward direction, the momentum of the airplane will
    * a. decrease.
      b. be unchanged.
      c. increase.

31. We can explain the recoil that occurs when a rifle is fired by using
      a. conservation of momentum.
      b. equal and opposite impulses.
      c. equal and opposite changes in momentum
    * d. Any of the above.

32. We can explain the recoil that occurs when a rifle is fired by using the conservation of
    * a. momentum.
      b. kinetic energy.
      c. work.
      d. mass.

33. What happens to the total momentum of a star that undergoes a supernova explosion?
      a. It increases.
    * b. It remains constant.
      c. It decreases.
      d. It depends on the color of the star.

34. When a star undergoes a supernova explosion, the total linear momentum of the star
    * a. does not change.
      b. increases.
      c. decreases.

35. An astronaut floating in the center of the main room in SkyLab with no translational motion relative to Skylab
      a. can get to a wall by waving his arms and legs.
      b. can convert his rotational motion to translational motion.
    * c. is stranded.
      d. cannot change his orientation.

36. Greg runs across the front of the classroom with a momentum of 300 kg·m/s and jumps onto a giant skateboard. The skateboard is initially at rest and has a mass equal to Greg's. If we ignore friction with the floor, what is the total momentum of Greg and the skateboard?
      a. zero
      b. 150 kg·m/s
    * c. 300 kg·m/s
      d. 600 kg·m/s

Chapter 6   Momentum

37. Carmen is standing on a giant skateboard that is initially at rest. If we ignore frictional effects with the floor, what is the momentum of the skateboard if Carmon walks to the right with a momentum of 500 kg·m/s?
    a. 500 kg·m/s to the right
    *   b. 500 kg·m/s to the left
    c. 500 kg·m/s
    d. zero

38. Two air-track gliders are held together with a string. The mass of glider A is twice that of glider B. A spring is tightly compressed between the gliders. If the gliders are initially at rest and the spring is released by burning the string, what is the total momentum of both gliders after the release?
    a. twice the momentum of A
    b. half the momentum of A
    c. twice the momentum of B
    *   d. zero

39. Larry has a mass of 90 kg and runs across the front of the classroom with a speed of 5 m/s and jumps onto a giant skateboard. The skateboard is initially at rest and has a mass equal to Larry's. If we ignore friction with the floor, what is the speed of Larry and the skateboard?
    *   a. 2.5 m/s
    b. 5 m/s
    c. 45 m/s
    d. 450 m/s

40. Juan has a mass of 60 kg and is standing on a giant skateboard that is initially at rest and has a mass of 30 kg. If we ignore frictional effects with the floor, what is the speed of the skateboard if Juan walks to the right with a speed of 3 m/s?
    a. 1 m/s
    b. 3 m/s
    *   c. 6 m/s
    d. 600 m/s

41. Two air-track gliders are held together with a string. The mass of glider A is twice that of glider B. A spring is tightly compressed between the gliders. The gliders are initially at rest and the spring is released by burning the string. If glider B has a speed of 3 m/s after the release, how fast will glider A be moving?
    a. 1 m/s
    *   b. 1.5 m/s
    c. 3 m/s
    d. 6 m/s

42. A 2-kg ball traveling to the right with a speed of 3 m/s collides with a 4-kg ball traveling to the left with a speed of 2 m/s. The total momentum of the two balls after the collision is _____ kg-m/s to the _____.
    a. 14 ... left
    b. 14 ... right
    c. 2 ... right
    * d. 2 ... left

43. A 2-kg ball traveling to the right with a speed of 4 m/s collides with a 4-kg ball traveling to the left with a speed of 2 m/s. The total momentum of the two balls after the collision is
    a. 16 kg·m/s to the left
    b. 16 kg·m/s to the right
    c. 2 kg·m/s to the right
    * d. zero

44. A 3-kg ball traveling to the right with a speed of 4 m/s overtook and collided with a 4-kg ball traveling in the same direction with a speed of 2 m/s. The total momentum after the collision was _____ kg·m/s to the right.
    a. 8
    b. 12
    * c. 20
    d. 96

45. A father (m = 90 kg) and son (m = 45 kg) are standing facing each other on a frozen pond. The son pushes on the father and finds himself moving backward at 3 m/s after they have separated. How fast is the father moving?
    a. 1 m/s
    b. 1.5 m/s
    c. 3 m/s
    * d. 6 m/s

46. A woman with a mass of 50 kg runs at a speed of 6 m/s and jumps onto a giant skateboard with a mass of 30 kg. What is the combined speed of the woman and the skateboard?
    * a. 3.75 m/s
    b. 4.74 m/s
    c. 6 m/s
    d. 10 m/s

Chapter 6   Momentum

47. A 3-kg ball traveling to the right with a speed of 4 m/s collides with a 4-kg ball traveling to the left with a speed of 3 m/s. What is the total momentum of the two balls before and after the collision?
    * a. zero
    b. 12 kg·m/s
    c. 24 kg·m/s
    d. The momentum will be different before and after the collision.

48. A 4-kg ball traveling to the right with a speed of 4 m/s collides with a 5-kg ball traveling to the left with a speed of 2 m/s. What is the total momentum of the two balls after they collide?
    * a. 6 kg·m/s to the right
    b. 22 kg·m/s to the right
    c. 26 kg·m/s to the right
    d. 40 kg·m/s to the right

49. A 1200-kg car traveling north at 14 m/s is rear-ended by a 2000-kg truck traveling at 25 m/s. What is the total momentum before and after the collision?
    a. 16,800 kg·m/s
    b. 33,200 kg·m/s
    c. 62,400 kg·m/s
    * d. 66,800 kg·m/s

50. A 1200-kg car traveling north at 14 m/s is rear-ended by a 2000-kg truck traveling at 25 m/s. If the truck and car lock bumpers and stick together, what is their speed immediately after the collision?
    a. 5.25 m/s
    b. 10.4 m/s
    c. 19.5 m/s
    * d. 20.9 m/s

51. A boxcar traveling at 10 m/s approaches a string of three identical boxcars sitting stationary on the track. The moving boxcar collides and links with the stationary cars, and the four move off together along the track. What is the final speed of the four cars immediately after the collision? (You may take the mass of each boxcar to be 18,537 kg.)
    a. 2 m/s
    * b. 2.5 m/s
    c. 3.33 m/s
    d. 10 m/s

Chapter 7: Energy

1. Under what conditions is energy conserved?
    a. When the system is closed to the flow of matter.
    b. When the system is closed to the flow of energy.
    * c. When the system is closed to the flow of matter and energy.
    d. Never

2. Which of the following physical quantities is not necessarily an invariant even if the system is closed?
    a. mass
    b. energy
    c. momentum
    * d. volume

3. When a star undergoes a supernova explosion, the total energy of the universe
    * a. does not change.
    b. increases.
    c. decreases.

4. Kinetic energy is defined to be one-half the
    a. mass times the speed.
    * b. mass times the speed squared.
    c. mass times the acceleration.
    d. weight times the speed squared.

5. Which has the greater kinetic energy, a heavy truck at rest or a moving roller skate?
    a. the heavy truck
    * b. the roller skate
    c. The kinetic energies are equal.

6. Which has the greater kinetic energy, an 18-wheeler parked at the curb or a Volkswagen rolling down a hill?
    a. 18-wheeler
    * b. Volkswagen
    c. The kinetic energies are equal.

7. If a sports car with a mass of 1000 kg travels down the road with a speed of 20 m/s, its kinetic energy is 200,000
    a. $kg/(m/s)^2$
    b. $kg \cdot s^2/m^2$
    c. $kg \cdot m/s$
    * d. $kg \cdot m^2/s^2$

Chapter 7   Energy

8. A sports car with a mass of 1000 kg travels down the road with a speed of 20 m/s. Which is larger, its momentum or its kinetic energy?
    a. momentum
    b. kinetic energy
    c. They are equal.
*   d. The two cannot be compared.

9. Which of the following objects has the largest kinetic energy? A mass of ____ with a speed of _____.
    a. 6 kg ... 2 m/s
    b. 4 kg ... 3 m/s
    c. 2 kg ... 5 m/s
*   d. 6 kg ... 3 m/s

10. If a sports car with a mass of 1000 kg travels down the road with a speed of 30 m/s, its kinetic energy is
    a. 15,000 J.
    b. 30,000 J.
*   c. 450,000 J.
    d. 900,000 J.

11. What is the kinetic energy of an 80-kg sprinter running at 9 m/s?
    a. 360 J
    b. 720 J
*   c. 3240 J
    d. 6480 J

12. Assume that a red car has a mass of 1000 kg and a white car has a mass of 2000 kg. If both cars are traveling at the same velocity,
    a. their kinetic energies are equal.
    b. their momenta are equal.
    c. the red car's kinetic energy is twice as big.
*   d. the white car's kinetic energy is twice as big.

13. Assume that a red car has a mass of 1000 kg and a white car has a mass of 2000 kg. If the red car has twice the velocity of the white car,
    a. their kinetic energies are equal.
    b. their momenta are equal.
*   c. the red car's momentum is twice as big.
    d. the white car's momentum is twice as big.

14. Assume that two cars have the same mass, but that the red car has twice the speed of the blue car. We then know that the red car has _____ kinetic energy as the blue car.
    a. twice as much
    b. one-half as much
  * c. four times as much
    d. one-fourth as much

15. Under what conditions is the kinetic energy conserved in a collision?
    a. It is always conserved.
  * b. When the collision is totally elastic.
    c. When there is no net outside force.
    d. When there is no friction.

16. A ball moving at 4 m/s toward the right has a head-on collision with an identical stationary ball. Each of the following possibilities satisfies the law of conservation of linear momentum. Which one also conserves kinetic energy? One ball has a velocity of _____ while the other has a velocity of _____ to the right.
    a. 2 m/s to the right ... 2 m/s
  * b. zero ... 4 m/s
    c. 2 m/s to the left ... 6 m/s
    d. 4 m/s to the left ... 8 m/s

17. A ball moving at 4 m/s toward the right has a head-on collision with an identical stationary ball. Each of the following possibilities satisfies the law of conservation of linear momentum. In which one is the most kinetic energy lost? One ball has a velocity of _____ while the other has a velocity of _____ to the right.
  * a. 2 m/s to the right ... 2 m/s
    b. zero ... 4 m/s
    c. 2 m/s to the left ... 6 m/s
    d. 4 m/s to the left ... 8 m/s

18. A 3-kg toy car with a speed of 6 m/s collides head-on with a 2-kg car traveling in the opposite direction with a speed of 4 m/s. If the cars are locked together after the collision with a speed of 2 m/s, how much kinetic energy is lost?
    a. 10 J
    b. 28 J
  * c. 60 J
    d. 70 J

19. A 4-kg toy car with a speed of 5 m/s collides head-on with a stationary 1-kg car. After the collision, the cars are locked together with a speed of 4 m/s. How much kinetic energy is lost in the collision?
    * a. 10 J
      b. 40 J
      c. 50 J
      d. 90 J

20. In physics, work is defined as the product of the
      a. net force and the distance traveled.
    * b. net force parallel to the motion and the distance traveled.
      c. net force parallel to the motion and the time it is applied.
      d. applied force and the distance traveled.

21. An object has a velocity toward the south. If a force directed toward the south acts on the object, it will initially cause the kinetic energy of the object to _____.
    * a. increase
      b. decrease
      c. remain the same

22. An object has a velocity toward the south. If a force directed toward the north acts on the object, it will initially cause the kinetic energy of the object to _____.
      a. increase
    * b. decrease
      c. remain the same

23. Two objects have different masses but the same kinetic energies. If you stop them with the same retarding force, which one will stop in the shorter distance?
      a. the heavier one
      b. the lighter one
    * c. Both stop in the same distance.

24. Two objects have different masses but the same momenta. If you stop them with the same retarding force, which one will stop in the shorter distance?
    * a. the heavier one
      b. the lighter one
      c. Both stop in the same distance.

25. The kinetic energy of an object moving in a circle at a constant speed is
    a. continually changing as the force changes direction.
    b. equal to the force times the time for one revolution.
    c. equal to one-half of the potential energy.
    *   d. constant.

26. Which of the following forces does the most work? A force of _____ acting through a distance of _____.
    a. 1 N ... 5 m
    b. 2 N ... 4 m
    *   c. 3 N ... 3 m
    d. 4 N ... 2 m

27. A bucket of water is raised from a well using a rope wrapped around a cylinder. The cylinder is turned by a crank that is 48 cm long. The handle follows a circle that has a circumference of 3 m. If the force on the crank is 200 N, how much work is done each time it goes around?
    a. zero
    b. 248 J
    *   c. 600 J
    d. 9600 J

28. A tennis ball on the end of a string travels in a horizontal circle at a constant speed. The circle has a circumference of 2 m, the ball has a speed of 3 m/s, and the centripetal force is 1.5 N. How much work is done on the ball each time it goes around?
    *   a. zero
    b. 6 J
    c. 9 J
    d. 12 J

29. How much work is performed by the gravitational force F on a synchronous satellite during one day?
    a. zero, because the satellite does not move.
    *   b. zero, because the force is perpendicular to the velocity.
    c. FC, where C is the circumference of the orbit.
    d. Fr, where r is the radius of the orbit.

30. A bowler lifts a bowling ball from the floor and places it on a rack. If you know the weight of the ball, what else must you know in order to calculate the work she does on the ball?
    a. the mass of the ball
    b. the time required
    c. nothing else
    *   d. the height of the rack

31. What is the gravitational potential energy of a ball with a weight of 50 N when it is sitting on a shelf 2 m above the floor?
     a. 100 J
     b. 200 J
     c. 400 J
   * d. We cannot say without knowing the zero level.

32. Which one of the following does not affect the value for the gravitational potential energy of an object?
     a. its height
     b. the location of the zero value
     c. its mass
   * d. All of them do.

33. A woman with a mass of 60 kg climbs a set of stairs that are 3 m high. How much gravitational potential energy does she gain?
     a. 60 J
     b. 180 J
     c. 540 J
   * d. 1800 J

34. How much work does a 60-kg person do against gravity in walking up a trail that gains 720 m in elevation?
     a. 12 J
     b. 780 J
     c. 43,200 J
   * d. 432,000 J

35. What is the gravitational potential energy of a ball with a weight of 50 N when it is sitting on a shelf 2 m above the floor? Assume the potential energy is zero on the floor.
   * a. 100 J
     b. 200 J
     c. 400 J
     d. zero

36. A man with a mass of 70 kg falls 10 m. How much gravitational potential energy does he lose?
     a. 10 J
     b. 70 J
     c. 700 J
   * d. 7000 J

37. Which of the following properties of a ball is conserved as it falls freely in a vacuum?
    a. kinetic energy
    b. gravitational potential energy
    c. momentum
  * d. mechanical energy

38. If we examine a ball in free fall, we find that the kinetic energy of the ball is not constant. This is not a violation of the law of conservation of energy because the
    a. force of gravity does work on the ball.
    b. system is not closed.
    c. the gravitational potential energy also changes.
  * d. All of the above are correct.

39. A man with a mass of 70 kg falls 10 m. How much kinetic energy does he gain?
    a. 10 J
    b. 70 J
    c. 700 J
  * d. 7000 J

40. If a 0.5-kg ball is dropped from a height of 6 m, what is its kinetic energy when it hits the ground?
    a. 3 J
    b. 9 J
  * c. 30 J
    d. There is not enough information to say.

41. A 1200-kg frictionless roller coaster starts from rest at a height of 20 m. What is its kinetic energy when it goes over a hill that is 10 m high?
    a. 12,400 J
    b. 24,000 J
  * c. 120,000 J
    d. 240,000 J

42. A ball dropped from a height of 10 m only bounces to a height of 5 m. Which of the following statements is valid for this situation?
    a. Kinetic energy is conserved.
    b. Mechanical energy is conserved.
    c. Gravitational potential energy is conserved.
  * d. None of the above.

43. At which point in the swing of an ideal pendulum (ignoring friction) is the kinetic energy a maximum?
    a. at either end
    * b. at the lowest point
    c. It's always the same.

44. An elephant, an ant, and a professor jump from a lecture table. Assuming no frictional losses, which of the following could be said just before they hit the floor?
    a. They all have the same kinetic energy.
    b. They all started with the same potential energy.
    c. They will all experience the same force on stopping.
    * d. They all have the same speed.

45. A 1-kg ball falling freely through a distance of one meter loses 10 J of gravitational potential energy. How much does the kinetic energy of the ball change if this occurs in a vacuum?
    a. gain of 1 J
    * b. gain of 10 J
    c. loss of 1 J
    d. loss of 10 J

46. A 3-kg mass is released at the top of a frictionless slide that is 2 m high. What is the kinetic energy of the mass when it reaches the bottom?
    a. 9 J
    b. 18 J
    c. 30 J
    * d. 60 J

47. A block of wood loses 160 J of gravitational potential energy as it slides down a ramp. If it has 90 J of kinetic energy at the bottom of the ramp, we can conclude that
    a. mechanical energy is conserved.
    b. momentum is conserved.
    c. 250 J of energy was lost.
    * d. 70 J of energy was transformed to another form.

48. A 0.5-kg air-hockey puck is initially at rest. What will its kinetic energy be after a net force of 0.6 N acts on it for a distance of 2 m?
    a. 0.3 J
    b. 0.6 J
    * c. 1.2 J
    d. 2.4 J

49. A 20-N block lifted straight upward by a hand applying a force of 20 N has an initial kinetic energy of 26 J. If the block is lifted 1 m, what is the block's final kinetic energy?
    - a. 6 J
* - b. 26 J
    - c. 40 J
    - d. 46 J

50. A radio-controlled car increases its kinetic energy from 4 J to 12 J over a distance of 2 m. What was the average net force on the car during this interval?
* - a. 4 N
    - b. 6 N
    - c. 8 N
    - d. 12 N

51. A toy car has a kinetic energy of 12 J. What is its kinetic energy after a frictional force of 0.6 N has acted on it for 5 m?
    - a. 3 J
    - b. 4 J
    - c. 6 J
* - d. 9 J

52. Which of the following is NOT a potential energy?
    - a. elastic
* - b. friction
    - c. chemical
    - d. nuclear

53. Which of the following is NOT a unit of energy?
    - a. joule
    - b. newton-meter
    - c. kilowatt-hour
* - d. watt

54. Which of the following is an energy unit?
    - a. newton
    - b. kilowatt
    - c. kilogram-meter/second
* - d. kilowatt-hour

Chapter 7   Energy

55. Power is defined to be the energy
      a. lost in a process.
      b. lost in a process divided by the time it takes.
      c. changed to other forms in a process.
  * d. changed to other forms divided by the time it takes.

56. If a 60-kg sprinter can accelerate from a standing start to a speed of 10 m/s in 2 s, what average power is generated?
      a. 600 W
      b. 1200 W
  * c. 1500 W
      d. 3000 W

57. What average power is required to accelerate a 1200-kg car from rest to 20 m/s in 10 s?
      a. 240 W
  * b. 24,000 W
      c. 36,000 W
      d. 48,000 W

58. How much energy is required to light a 100-W bulb for 5 h?
      a. 2500 W-h
  * b. 500 W-h
      c. 100 W-h
      d. 20 W-h

59. If a CD player uses electricity at a rate of 15 W, how much energy does it use during an 8-h day?
      a. 15 J
      b. 120 J
      c. 720 J
  * d. 432,000 J

60. If a hair dryer is rated at 1500 W, how much energy does it require in 5 min?
      a. 300 J
      b. 1500 J
      c. 7500 J
  * d. 450,000 J

Chapter 8: Rotation

1. If the beaters on a mixer make 800 revolutions in 5 minutes, what is the average rotational speed of the beaters?
   a. 2.67 rev/min
   b. 16.8 rev/min
   * c. 160 rev/min
   d. 4000 rev/min

2. What is the rotational speed of the hand on a clock that measures the minutes?
   a. 0.000278 rad/s
   * b. 0.00175 rad/s
   c. 0.0167 rad/s
   d. 0.105 rad/s

3. What is the rotational speed of the hand on a clock that measures the minutes?
   * a. 0.000278 rev/s
   b. 0.00175 rev/s
   c. 0.0167 rev/s
   d. 0.105 rev/s

4. What is the rotational speed of the hand on a clock that measures the seconds?
   a. 0.000278 rev/s
   b. 0.00175 rev/s
   * c. 0.0167 rev/s
   d. 0.105 rev/s

5. What is the rotational speed of the second hand on a clock that measures the seconds?
   a. 0.000278 rad/s
   b. 0.00175 rad/s
   c. 0.0167 rad/s
   * d. 0.105 rad/s

6. If a phonograph turntable takes 2 s to reach its rotational speed of 45 rpm, what is its average acceleration?
   a. 2 rpm/s
   b. 11.2 rpm/s
   * c. 22.5 rpm/s
   d. 45 rpm/s

Chapter 8   Rotation

7. If it takes 3 s for a modern player to stop a DVD with a rotational speed of 7500 rpm, what is the magnitude of the DVD's average rotational acceleration?
    * a. 2500 rpm/s
    b. 22,500 rpm/s
    c. 150,000 rpm/s
    d. 1,350,000 rpm/s

8. If it takes 3 s for a modern player to stop a DVD with a rotational speed of 7500 rpm, what is the magnitude of the DVD's average rotational acceleration?
    a. 2500 rev/s^2
    b. 22,500 rev/s^2
    * c. 150,000 rev/s^2
    d. 1,350,000 rev/s^2

9. If the wheel on a car accelerates from rest at a constant rate of 1.5 rev/s^2, how fast will it be turning after 3 s?
    a. 1.5 rev/s
    b. 2 rev/s
    c. 3 rev/s
    * d. 4.5 rev/s

10. A change in the rotational velocity of an object is produced by a net
    a. force.
    * b. torque.
    c. rotational inertia.
    d. moment arm.

11. Newton's first law for rotational motion states that an object will maintain its state of rotational motion unless acted on by an unbalanced (or net)
    a. force.
    b. velocity.
    c. inertia.
    * d. torque.

12. Newton's first and second laws for translational motion refer to the net force acting on an object. What is the rotational analog of this force?
    * a. torque
    b. inertia
    c. acceleration
    d. force

13. What kind of motion does a constant, non-zero torque produce on an object mounted on an axle?
      a. constant rotational speed
*   b. constant rotational acceleration
      c. increasing rotational acceleration
      d. decreasing rotational acceleration

14. If a ball with a weight of 30 N hangs from the end of a 1.5-m horizontal pole, what torque does the ball exert?
      a. 20 N·m
      b. 30 N·m
*   c. 45 N·m
      d. 60 N·m

15. A pirate with a mass of 90 kg stands on the end of a plank that extends 2 m beyond the gunwale. What torque is needed to keep him from falling into the water?
      a. 45 N·m
      b. 180 N·m
      c. 450 N·m
*   d. 1800 N·m

16. You are holding a 6-kg dumbbell straight out at arm's length. Assuming your arm is 0.70 m long, what torque is the dumbbell exerting on your shoulder?
      a. 4.2 N·m
      b. 8.6 N·m
*   c. 42 N·m
      d. 86 N·m

17. A meter stick is balanced on a stand. If you place a 50-g mass on one side at a distance of 20 cm from the center, how far from the center would you place a 25-g mass on the other side so that the system balances?
      a. 10 cm
      b. 20 cm
      c. 30 cm
*   d. 40 cm

18. Two children with masses of 20 and 30 kg are sitting on a balanced seesaw. If the lighter child is sitting 3 m from the center, how far from the center is the heavier child sitting?
      a. 1 m
*   b. 2 m
      c. 3 m
      d. 5 m

Chapter 8   Rotation

19. A child with a mass of 20 kg sits at a distance of 2 m from the pivot point of a seesaw. Where should a 16-kg child sit to balance the seesaw?
     a. 1.6 m
     b. 2.0 m
   * c. 2.5 m
     d. 3.2 m

20. An object's resistance to a change in its rotational velocity is called its
     a. inertia.
     b. mass.
     c. torque.
   * d. rotational inertia.

21. The rotational inertia of an object increases as the mass _____ and the distance of the mass from the center of rotation _____.
   * a. increases ... increases
     b. increases ... decreases
     c. decreases ... increases
     d. decreases ... decreases

22. A solid disk and a hoop have the same mass and radius. Which has the larger rotational inertia about its center of mass?
   * a. hoop
     b. disk
     c. They are the same.

23. A solid sphere and a solid cylinder are made of the same material. If they have the same mass and radius, which one has the larger rotational inertia about its center?
     a. the sphere
   * b. the cylinder
     c. They are the same.

24. Two flywheels have the same mass but one is much thinner than the other such that its radius is twice that of the smaller one. If both flywheels are spinning about their axes at the same rate, which one would be harder to stop?
   * a. the one with the larger radius
     b. the one with the smaller radius
     c. They are equally hard to stop.

25. In which of the following positions would a diver have the smallest rotational inertia for performing a front somersault?
    *   a. tuck
        b. pike
        c. layout

26. When an Olympic caliber high-jumper executes the winning jump, we find that her center of mass
        a. must pass over the top of the bar by one-half the width of her body.
        b. must just clear the bar.
    *   c. passes below the bar.

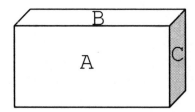

27. The block of wood shown above can be laid face down on any of the three marked faces. Face B is narrower than face A. Which would you put face down to get the least stable position?
        a. A
        b. B
    *   c. C

28. A fully-loaded trailer truck is less stable than a race car because the truck
        a. is more massive.
    *   b. has a higher center of mass.
        c. has bigger wheels.
        d. weighs more.

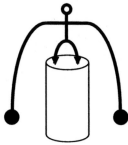

29. The balancing child on the pedestal in the diagram above is able to stay upright because the center of mass is
    *   a. located below the top of the pedestal.
        b. located at the knees of the child.
        c. located at the center of the child.
        d. not important. It is the total mass that determines whether the child stays upright.

30. If you stand with your back against a wall, you find that you cannot bend over to pick something up because your
    * a. center of mass moves forward beyond your toes.
      b. vertical and horizontal motions are independent.
      c. center of mass is located at a fixed position in your body.
      d. center of mass hits the wall.

31. If you face a wall with your toes against the baseboard, you find that you cannot stand up on your toes because
      a. you cannot generate the necessary torque.
      b. your rotational inertia is too large.
      c. your center of mass is too high.
    * d. your center of mass cannot move forward over your toes.

32. A child with a mass of 25 kg is riding on a merry-go-round. If the child has a speed of 3 m/s and is located 2 m from the center of the merry-go-round, what is the child's angular momentum?
      a. 50 kg·m^2/s
      b. 75 kg·m^2/s
    * c. 150 kg·m^2/s
      d. 300 kg·m^2/s

33. A 1800-kg car is traveling at 30 m/s around a curve with a radius of 120 m. What is the angular momentum of the car?
      a. 450 kg·m^2/s
      b. 7,200 kg·m^2/s
      c. 54,000 kg·m^2/s
    * d. 6,480,000 kg·m^2/s

34. Under what conditions is the total angular momentum of a system conserved?
      a. It is always conserved.
      b. When there is no net outside force.
    * c. When there is no net outside torque.
      d. When the kinetic energy is also conserved.

35. The reason a figure skater spins slower as he extends his arms is because
      a. he experiences a smaller torque.
      b. his angular momentum is greater.
      c. his angular momentum is less.
    * d. his rotational inertia is greater.

36. An astronaut "floating" in Skylab has an initial rotational motion but no initial translational motion relative to Skylab. She continues to rotate
    a. because she experiences no net force.
    * b. because the net force acts through her center of mass.
    c. because she is weightless.
    d. because she experiences a torque due to the force of gravity.

37. It is possible for a high diver to execute more front somersaults in the tuck position than in the layout position because _____ in the tuck position.
    a. her angular momentum is greater
    * b. her rotational inertia is less
    c. she exerts a greater torque
    d. her linear momentum is greater

38. When a gymnast performs a double somersault with a full twist in the middle, the angular momentum for the twist
    a. comes from the torque due to gravity.
    * b. comes from the somersaulting motion.
    c. comes from the force of gravity.

39. If Earth's orbit were circular, each of the following physical quantities would be conserved. For an elliptical orbit only _____ is conserved.
    a. speed
    * b. angular momentum
    c. kinetic energy
    d. size of the linear momentum

40. A cat which is held upside down and dropped with no initial angular momentum manages to land on its feet. Where does the cat get the necessary angular momentum?
    a. from the air resistance
    b. from gravity
    c. from the torque
    * d. It doesn't, none is needed.

41. A gyroscope which points directly up when it is located at the North Pole is transported to the equator while it is still spinning. Which way will it point?
    a. up
    b. down
    * c. north
    d. south

Chapter 8   Rotation

42. A spinning gyroscope points directly at the North Star when it is located at the North Pole. If it is transported to the South Pole without exerting any torques on it, which way will it point?
      a. up
*   b. down
      c. horizontally

43. Assume that Gerry sits on a freely rotating stool holding a bicycle wheel with its axle vertical so that it rotates in a clockwise direction when viewed from above. If Gerry turns the wheel over, he will
      a. not rotate because the system of wheel and Gerry is closed or isolated.
      b. not rotate because the two torques cancel.
*   c. rotate clockwise because angular momentum is conserved.
      d. rotate counterclockwise because angular momentum is conserved.

44. A spinning top precesses because
      a. angular momentum is conserved.
      b. it experiences no net torque.
      c. it experiences no net force.
*   d. gravity exerts a torque on it.

Chapter 9: Classical Relativity

1. Imagine riding in a glass-walled elevator that goes up the outside of a tall building at a constant speed of 20 meters per second. Assuming that you drop a ball, you will observe the ball
   * a. fall starting from rest.
   b. fall starting with an upward speed of 20 m/s.
   c. fall starting with a downward speed of 20 m/s.
   d. remain stationary.

2. Imagine riding in a glass-walled elevator that goes up the outside of a tall building at a constant speed of 20 meters per second. Assuming that you drop a ball as you pass a window washer, the window washer will see the ball
   a. fall starting from rest.
   * b. fall starting with an upward speed of 20 m/s.
   c. fall starting with a downward speed of 20 m/s.
   d. remain stationary.

3. Imagine riding in a glass-walled elevator that goes up the outside of a tall building at a constant speed of 20 meters per second. If a window washer drops a ball as you pass by, you will observe the ball
   a. fall starting from rest.
   b. fall starting with an upward speed of 20 m/s.
   * c. fall starting with a downward speed of 20 m/s.
   d. remain stationary.

4. While you are standing on the ground, you observe your friends pass by in a van traveling at a constant velocity. They drop a ball and you all describe its motion accurately. Which description is correct?
   a. yours
   b. your friends'
   * c. both
   d. neither

5. While you are standing on the ground, you observe your friends pass by in a van traveling at a constant velocity. They drop a ball and you all accurately describe the forces that are involved. Which description is correct?
   a. yours
   b. your friends'
   * c. both
   d. neither

6. While you are standing on the ground, you observe your friends pass by in a van traveling at a constant velocity. They drop a ball and you all accurately describe the velocity of the ball. Which description is correct?

        a. yours

        b. your friends'

    *   c. both

        d. neither

7. While you are standing on the ground, you observe your friends pass by in a van traveling at a constant velocity. They drop a ball and you all make measurements of the ball's motion. Which of the following quantities has the same value in both reference systems?

        a. velocities

        b. kinetic energies

    *   c. forces

        d. momenta

8. A reference system in which the law of inertia is valid is known as _____ reference system.

        a. a centripetal

        b. a circumstantial

    *   c. an inertial

        d. an accelerating

9. An inertial system is defined as one in which Newton's _____ law of motion is valid.

    *   a. first

        b. second

        c. third

10. Any reference system moving with a constant velocity with respect to an inertial system is known as _____ reference system.

        a. an alternate

        b. a noninertial

    *   c. an inertial

        d. an accelerating

11. Assume that you are riding in a windowless room on a perfectly smooth surface. (You can't feel any motion.) Imagine that you have a collection of objects and measuring devices in the room. Which of the following experiments could you do to prove that the room is moving horizontally at a constant velocity?

        a. Determining an object's mass by applying a net horizontal force.

        b. Weighing an object and comparing it to its known weight.

        c. Determining the force necessary for an object to move in a circle.

    *   d. None of the above.

12. You can throw a ball vertically up in a car moving with a constant velocity and have it land back in your hand because
    * a. there is no net horizontal force acting on the ball.
    b. the reference system attached to the car is noninertial.
    c. there is a net force in the forward direction.
    d. the force in the forward direction is canceled by the inertial force.

13. A train is traveling along a straight, horizontal track at a constant velocity of 50 mph. An observer in the train holds a ball directly over a white spot on the floor of the train. If she drops the ball, it will land ____ the white spot.
    a. behind
    * b. on
    c. in front of
    d. to the side of

14. A person drops a ball in train traveling along a straight, horizontal track at a constant velocity of 50 mph. An observer on the ground (that is, standing next to the tracks) would determine that the horizontal velocity of the ball during the fall is
    a. zero
    * b. equal to 50 mph
    c. increasing
    d. decreasing

15. A person drops a ball in a train traveling along a straight, horizontal track at a constant velocity. What would the observer on the ground say about the horizontal forces acting on the ball?
    * a. There are no horizontal forces acting on the ball.
    b. There is a fictitious (inertial) force acting forward.
    c. There is a fictitious (inertial) force acting backward.
    d. There is a centrifugal force.

16. A person drops a ball in a train traveling along a straight, horizontal track at a constant velocity. What would the person in the train say about the horizontal forces acting on the ball?
    * a. There are no horizontal forces acting on the ball.
    b. There is a fictitious (inertial) force acting forward.
    c. There is a fictitious (inertial) force acting backward.
    d. There is a centrifugal force.

17. A person drops a ball in a train traveling along a straight, horizontal track at a constant velocity. An observer on the ground would determine that the ball's acceleration is _____ *g*.
    a. greater than
    * b. equal to
    c. less than

18. A person drops a ball in a train traveling along a straight, horizontal track at a constant velocity. The person in the train would determine that the ball's acceleration is _____ g.
       a. greater than
    * b. equal to
       c. less than

19. A train is traveling along a straight, horizontal track with a constant speed of 50 mph. If the ball is thrown forward with a speed of 60 mph relative to the train, what is its speed relative to the ground?
       a. 10 mph
       b. 50 mph
       c. 60 mph
    * d. 110 mph

20. A train is traveling along a straight, horizontal track with a constant speed of 50 mph. If the ball is thrown backward with a speed of 60 mph relative to the train, what is its speed relative to the ground?
    * a. 10 mph
       b. 50 mph
       c. 60 mph
       d. 110 mph

21. A ball is thrown horizontally at 30 m/s from a flatcar that is moving in a straight line at 40 m/s. Relative to a person on the ground, what is the horizontal speed of the ball when it is thrown in the forward direction?
       a. 10 m/s
       b. 30 m/s
       c. 50 m/s
    * d. 70 m/s

22. A rock is thrown horizontally at 10 m/s from the back of a flatbed truck that is moving with a constant velocity of 30 m/s. Relative to an observer on the ground, what is the horizontal velocity of the rock when it is thrown in the backward direction?
    * a. 20 m/s forward
       b. 20 m/s backward
       c. 40 m/s forward
       d. 40 m/s backward

23. A ball is thrown horizontally at 30 m/s from a flatcar that is moving in a straight line at 40 m/s. Relative to a person on the ground, what is the horizontal speed of the ball when it is thrown directly sideways?
    a. 10 m/s
    b. 30 m/s
    * c. 50 m/s
    d. 70 m/s

24. An aircraft carrier is moving to the north at a constant 25 mph on a windless day. A plane requires a speed relative to the air of 125 mph to take off. How fast must the plane be traveling relative to the deck of the aircraft carrier to take off if the plane is headed north?
    a. 25 mph
    * b. 100 mph
    c. 125 mph
    d. 150 mph

25. An aircraft carrier is moving to the north at a constant 25 mph on a windless day. A plane requires a speed relative to the air of 125 mph to take off. How fast must the plane be traveling relative to the deck of the aircraft carrier to take off if the plane is headed south?
    a. 25 mph
    b. 100 mph
    c. 125 mph
    * d. 150 mph

26. A train is traveling along a straight, horizontal track with a constant acceleration in the forward direction. At the instant the speed is 50 mph, a ball is dropped by an observer in the train. An observer on the ground determines that the horizontal speed of the ball during the fall is
    a. decreasing.
    b. increasing.
    c. zero.
    * d. equal to 50 mph.

27. A train is traveling along a straight, horizontal track with a constant acceleration in the forward direction. At the instant the speed is 50 mph, a ball is dropped by an observer in the train. The observer in the train determines that the horizontal speed of the ball during the fall is
    a. decreasing.
    * b. increasing.
    c. zero.
    d. equal to 50 mph.

Chapter 9   Classical Relativity

28. An observer drops a ball in a train traveling along a straight, horizontal track with a constant acceleration in the forward direction. The observer in the train determines that the vertical acceleration of the ball is _____ g.
    a. greater than
* b. equal to
    c. less than

29. An observer drops a ball in a train traveling along a straight, horizontal track with a constant acceleration in the forward direction. An observer on the ground determines that the vertical acceleration of the ball is _____ g.
    a. greater than
* b. equal to
    c. less than

30. An observer drops a ball in a train traveling along a straight, horizontal track with a constant acceleration in the forward direction. What would an observer in the train say about the horizontal force acting on the ball?
    a. There is no horizontal force.
* b. A force acts backward.
    c. A force acts forward.
    d. There is a centrifugal force.

31. An observer drops a ball in a train traveling along a straight, horizontal track with a constant acceleration in the forward direction. What would an observer on the ground say about the horizontal force acting on the ball?
* a. There is no horizontal force.
    b. A fictitious force acts backward.
    c. A fictitious force acts forward.
    d. There is a centrifugal force.

32. An observer drops a ball in a train traveling along a straight, horizontal track with a constant acceleration in the forward direction. The observer is unaware of the acceleration and notices that the ball falls in a straight line that is slanted toward the back of the train. The acceleration of the ball along this line is _____ g.
* a. greater than
    b. equal to
    c. less than

33. You and a friend are rolling marbles on a horizontal table in the back of a moving van on a straight, level section of interstate highway. You start the marble rolling directly toward the side of the truck and observe that it curves toward the front. You conclude that the truck is
    a. not moving
    b. moving at a constant velocity
    c. speeding up
    * d. slowing down

34. Clay stands on his bathroom scale in the morning and finds that he weighs 50 lbs. He takes a ride in the elevator that goes up the side of the Space Needle in Seattle. Much to the amusement of the other passengers he brings his bathroom scale and stands on it during the ride. During the time that the elevator is accelerating upward the reading on the scale is _____ 50 lbs.
    * a. greater than
    b. equal to
    c. less than

35. Rylan stands on his bathroom scale in the morning and finds that he weighs 40 lbs. He takes a ride in the elevator that goes up the side of the Space Needle in Seattle. Much to the amusement of the other passengers he brings his bathroom scale and stands on it during the ride. During the time that the elevator is accelerating downward the reading on the scale is _____ 40 lbs.
    * a. greater than
    b. equal to
    c. less than

36. Mehmet stands on his bathroom scale in the morning and finds that he weighs 190 lbs. He takes a ride in the elevator that goes up the side of the Space Needle in Seattle. Much to the amusement of the other passengers he brings his bathroom scale and stands on it during the ride. During the time that the elevator is traveling upward with a constant speed the reading on the scales is _____ 190 lbs.
    a. greater than
    * b. equal to
    c. less than

37. An elevator is moving downward and slowing down with an acceleration equal to one-quarter that of gravity. If a person who weighs 800 N when at rest on Earth steps on a bathroom scale in this elevator, what will the scale read?
    a. 200 N
    b. 600 N
    c. 800 N
    * d. 1000 N

38. An elevator is moving upward and slowing down with an acceleration equal to one-quarter that of gravity. If a person who weighs 800 N when at rest on Earth steps on a bathroom scale in this elevator, what will the scale read?
    a. 200 N
 * b. 600 N
    c. 800 N
    d. 1000 N

39. What does the scale read if a 5-kg cat lies on a bathroom scale in an elevator accelerating downward at 0.2 g?
    a. 4 N
 * b. 40 N
    c. 50 N
    d. 60 N

40. As the Apollo astronauts passed the point in space where the gravitational attraction of Earth and the Moon on the spacecraft were equal,
    a. their sense of up and down reversed.
    b. they became weightless.
    c. they turned the spacecraft around so that they could still stand on the floor.
 * d. they started to speed up.

41. If a child weighs 200 N standing at rest on Earth, what is the weight of this child in an elevator being accelerated downward with a constant value of 4 m/s^2?
    a. zero
    b. 80 N
 * c. 120 N
    d. 200 N

42. An elevator is being accelerated upward with an acceleration equal to one-half that of gravity. If a man who weighs 160 lbs. when he is at rest on Earth steps on a bathroom scale in the elevator, the reading will be
    a. zero
    b. 80 lbs
    c. 160 lbs
 * d. 240 lbs.

43. What would an observer measure for the magnitude of the free-fall acceleration in an elevator near Earth's surface if the elevator accelerates downward at 6 m/s²?
    *   a. 4 m/s²
        b. 6 m/s²
        c. 10 m/s²
        d. 16 m/s²

44. What would an observer measure for the magnitude of the free-fall acceleration in an elevator near Earth's surface if the elevator accelerates upward at 6 m/s²?
        a. 4 m/s²
        b. 6 m/s²
        c. 10 m/s²
    *   d. 16 m/s²

45. A room is being accelerated through space at 10 m/s² relative to the "fixed stars." It is far away from any massive objects. If a man weighs 700 N when he is at rest on Earth, how much will he weigh in the room?
        a. zero
        b. 350 N
    *   c. 700 N
        d. 1400 N

46. A woman with a weight of 600 N on Earth is in a rocket ship accelerating through space a long way from any massive objects. If the acceleration is 20 m/s², what is the weight of the woman in the ship?
        a. zero
        b. 300 N
        c. 600 N
    *   d. 1200 N

47. A woman with a weight of 700 N on Earth is in a spacecraft accelerating through space a long way from any massive objects. If the acceleration is 4 m/s², what is her weight in the ship?
        a. zero
        b. 175 N
    *   c. 280 N
        d. 2800 N

Chapter 9   Classical Relativity

48. A child rides near the outer edge of a merry-go-round. Which way is "up" in the reference system attached to the merry-go-round?
    *   a. up and in toward the center
        b. down and in toward the center
        c. up and out from the center
        d. down and out from the center

49. A student hangs two pendula from the outer edge of a turntable rotating at 78 rpm. Which figure best illustrates the positions of the pendula?
        a. | |
    *   b. / \
        c. \ /

50. While driving to the movies you decide to take advantage of a sharp right-hand corner to slide your date over next to you. (Assume that the seat is frictionless.) From your point of view your date experiences a net force to the left, while a person standing on the roadway says your date experiences
        a. a net force to the right.
        b. a net force to the left.
    *   c. a net force forward.
        d. no net force.

51. A large cylindrical space ship is drifting through space and rotating about the axis of the cylinder. Which way is "up" for the occupants of the space ship?
    *   a. toward the axis
        b. away from the axis
        c. along the axis
        d. there is no "up" direction

52. A space station, far from any large masses, can be spun so that people inside the station feel the effects of an "artificial gravity." This works because
        a. the station is an inertial reference system.
    *   b. the people experience an inertial force.
        c. living in outer space doesn't require an "up" or "down" direction.
        d. gravity extends throughout space.

53. A person who weighs 600 N when at rest is riding in the rotating cylinder ride. The cylinder rotates fast enough to create an 800-N centrifugal force. What is the magnitude of the person's weight in the rotating reference frame?
     a. 600 N
     b. 800 N
  * c. 1000 N
     d. 1400 N

54. A cylindrical space station with a 40-m radius is rotating so that points on the walls have speeds of 20 m/s. What is the acceleration due to this artificial gravity at the walls?
     a. 0.5 m/s^2
     b. 2 m/s^2
  * c. 10 m/s^2
     d. 20 m/s^2

55. In the geocentric model of the universe the
     a. Earth is at the center of the universe.
     b. Earth does not move.
     c. stars go in circles around Earth each day.
  * d. All of the above are true.

56. In the heliocentric model of the universe the
     a. sun is at the center of the universe.
     b. Earth rotates on its axis once each day.
     c. Earth revolves around the sun once each year.
  * d. All of the above are true.

57. Which of the following was NOT cited as evidence that Earth moves?
     a. The plane of a pendulum rotates.
  * b. The sun rises and sets each day.
     c. Hurricanes turn counterclockwise in the Northern Hemisphere; clockwise in the Southern.
     d. The stars exhibit parallax.

58. Assuming that Earth is a perfect sphere and that the force of gravity is constant over the surface, your weight as determined by a bathroom scale at the equator would be _____ that at the North Pole. (Use the same scales at both locations.)
     a. the same as
     b. greater than
  * c. less than

59. If you look down on Earth from a satellite, you notice that hurricanes in the Northern Hemisphere rotate in a _____ direction.
      a. clockwise
  * b. counterclockwise
      c. northerly
      d. southerly

Chapter 10: Einstein's Relativity

1. According to the special theory of relativity, all laws of nature are the same in reference systems which _____ relative to an inertial system.
        a. have a constant acceleration
    *   b. move at a constant velocity
        c. move in ellipses
        d. move in circles at a constant speed

2. The effects that we have studied in special relativity occur when we look at phenomena in
        a. a single inertial system.
        b. a single noninertial system.
    *   c. two inertial systems.
        d. two noninertial systems.

3. The first postulate of special relativity
        a. says that there is no absolute space.
        b. is a reaffirmation of the Galilean principle of relativity.
        c. states that the laws of physics are the same in all inertial reference systems.
    *   d. All of these statements are true.

4. In his theory of special relativity, Einstein
        a. abandoned the Galilean principle of relativity.
        b. abandoned Maxwell's equations for electricity and magnetism.
    *   c. reconciled the apparent conflict between the Galilean principle of relativity and Maxwell's equations.
        d. postulated the existence of an absolute reference system.

5. A train is traveling along a straight, horizontal track at a constant speed that is only slightly less than that of light. If an observer in the train holds a ball directly over a white spot on the floor and drops it, the ball will land _____ the white spot.
        a. behind
        b. in front of
        c. to the side of
    *   d. on

6. A train is traveling along a straight, horizontal track at a constant speed that is only slightly less than that of light. If an observer in the train drops a ball and measures its acceleration, she will obtain a value
    *   a. equal to g
        b. greater than g
        c. less than g.

Chapter 10   Einstein's Relativity

7. A train is traveling along a straight, horizontal track at a constant speed that is only slightly less than that of light. If an observer in the train drops a ball and an observer on the ground measures its acceleration, she will obtain a value
    *   a. equal to *g*
         b. greater than *g*
         c. less than *g*

8. What was the purpose in postulating the existence of the ether?
         a. It determined which reference system was the absolute one.
         b. To account for the slowing of Earth in its annual journey around the Sun.
    *   c. To provide the medium through which light traveled.
         d. To account for the time difference measured in the Michelson-Morley experiments.

9. In the Michelson-Morley experiments two light beams are raced at right angles to each other. What were the results?
         a. The beam traveling along the direction of Earth's motion always won.
         b. The beam traveling along the direction of Earth's motion always lost.
    *   c. The races always ended in ties.
         d. The results depended on the season of the year.

10. The second postulate of special relativity states that the speed of light
    *   a. is a constant in a vacuum.
         b. is constant relative to the ether.
         c. depends on the motion of the source.
         d. depends on the motion of the receiver.

11. If you approach a light beacon while traveling at one-half the speed of light (0.5 *c*), you will measure the speed of light from the beacon to be
         a. 0.5 *c*
         b. 0.7 *c*
    *   c. *c*
         d. 1.5 *c*

12. As a space ship approaches you in outer space at 50% of the speed of light, its rotating beacon sends out a pulse of light. An observer in the ship measures the speed of the light leaving the ship to be
         a. 50 % of *c*
         b. 75% of *c*
    *   c. equal to *c*
         d. 150% of *c*

13. As a space ship approaches you in outer space at 50% of the speed of light, its rotating beacon sends out a pulse of light. You measure the speed of this light to be
    a. 50% of c
    b. 75% of c
* c. equal to c
    d. 150% of c

14. A train is traveling along a straight, horizontal track at a constant speed that is only slightly less than that of light. An observer in the train turns on a light in the caboose and measures the length of time it takes to get to the engine. (Assume that the light is traveling in a vacuum.) Knowing the length of the train, he is able to calculate the speed of light and obtains a speed
    a. less than c
* b. equal to c
    c. greater than c

15. A train is traveling along a straight, horizontal track at a constant speed that is only slightly less than that of light. An observer in the train turns on a light in the caboose. (Assume that the light travels to the engine in a vacuum.) An observer on the ground uses her instruments to measure the speed of this light and obtains a value
    a. less than c.
* b. equal to c.
    c. greater than c.

16. The speed of light is represented by the symbol c and has a value equal to
    a. 300 km/h.
    b. 300 km/s.
    c. 300,000 km/h.
* d. 300,000 km/s.

17. Two rockets are approaching a space station from opposite directions. Each is traveling at 90% of the speed of light as measured by an observer in the space station. How fast does each rocket approach the other as measured by observers in the rocket?
    a. 90% of c
* b. slightly less than c
    c. c
    d. 180% of c

18. Two rocket ships approach a space station at 80% of the speed of light. Each pilot observes the other ship approaching at _____ the speed of light.
    a. 64%
    b. 80%
* c. 98%
    d. 160%

19. Which of the following concepts is NOT a relative one, that is, observers in different inertial systems agree on the observations.
    a. simultaneity of events at separated locations
    b. rate at which clocks run
* c. simultaneity of events at a single location
    d. synchronization of clocks

20. A train is traveling along a straight, horizontal track at a constant speed that is only slightly less than that of light. An observer in the train determines that firecrackers go off simultaneously in the engine and in the caboose. An observer on the ground determines that the firecracker in the _____ went off first.
    a. engine
* b. caboose
    c. They went off simultaneously for both observers.

21. A train is traveling along a straight, horizontal track at a constant speed that is only slightly less than that of light. An observer on the ground determines that firecrackers go off simultaneously in the engine and in the caboose. An observer in the train determines that the firecracker in the _____ went off first.
* a. engine
    b. caboose
    c. They went off simultaneously for both observers.

22. A train is traveling along a straight, horizontal track at a constant speed that is only slightly less than that of light. An observer on the ground reports that as the midpoint of the train passes an observer on the ground, simultaneous flashes occurred in the engine and caboose. The train observer determines that the flash in the engine occurred _____ the one in the caboose.
* a. before
    b. at the same time as
    c. after

Physics: A World View, Sixth Edition by Larry Kirkpatrick and Gregory Francis

23. As a friend passed you at a very high speed, she reported that she simultaneously exploded a firecracker at each end of her skateboard. Which one exploded first from your point of view?
     a. the one at the front
*   b. the one at the back
     c. They exploded simultaneously.

24. Two lights on lampposts flash simultaneously as seen by observers on the ground. In which direction would you have to be moving in order to see the right-hand light flash first?
*   a. to the right
     b. to the left
     c. It is not possible.

25. A giant chicken in the sky at a height of 30 miles lays eggs. On the ground these eggs would hatch in 5 seconds. It is possible to get the eggs to the ground before they hatch if we transport them
     a. at a speed greater than that of light.
     b. at a speed equal to that of light.
*   c. at a speed less than, but close to that of light.
     d. It is impossible to do this.

26. Superman wants to travel back to his native Krypton for a visit, a distance of 3,000,000,000,000 meters. (It takes light 10,000 seconds to travel this distance.) If Superman can hold his breath for 1000 s and travel at any speed less than that of light, can he make it?
     a. Not unless he stops off for a breath on his way.
     b. Not unless he goes faster than light.
     c. Not without help from NASA.
*   d. Yes.

27. If a musician plays middle C on a clarinet while traveling at 85% of the speed of light in a rocket ship, passengers in the ship will hear
     a. a lower note.
*   b. middle C.
     c. a higher note.

28. A train is traveling along a straight, horizontal track at a constant speed that is only slightly less than that of light. A warning light on the ground flashes once each second. An observer in the train measures the time between flashes to be
*   a. greater than one second.
     b. one second.
     c. less than one second.

29. A train is traveling along a straight, horizontal track at a constant speed that is only slightly less than that of light. An observer in the train determines that clocks on the ground run _____ clocks in the train.
      a. at the same rate as
*   b. slower than
      c. faster than

30. How fast would a person have to travel in order to go backward in time?
      a. slightly less than c.
      b. equal to c.
      c. greater than c.
*   d. This is impossible.

31. In *A Connecticut Yankee in King Arthur's Court* Mark Twain chronicles the adventures of a New England craftsman who in 1879 is suddenly transported back in time to Camelot in the year 528. What does the special theory of relativity say about this possibility?
      a. This is possible only if he traveled at the speed of light.
      b. This is possible only if he traveled faster than the speed of light.
*   c. This is not possible.
      d. Time runs backwards for speeds greater than that of light.

32. In the TV series "The Voyagers" two heroes used a time device to travel back in time to observe (and preserve) various historical events such as Lindberg's flight across the Atlantic Ocean in the Spirit of St. Louis. What does the special theory of relativity say about this possibility?
      a. This is possible only if they traveled faster than the speed of light.
      b. This is possible only if they traveled at a speed a little less than that of light.
      c. This is theoretically possible, but not practical at the current time.
*   d. This is impossible to do.

33. On the average, a neutron at rest lives for 17 minutes before it decays. If the neutrons are moving, they will have _____ average life.
*   a. a longer
      b. a shorter
      c. the same

34. On the average a muon at rest lives for 2.2 microseconds before it decays into an electron and two neutrinos. If the muons are traveling at 90% of the speed of light, they will have _____ average life.
*   a. a longer
      b. a shorter
      c. the same

35. What is the approximate value of the relativistic adjustment factor for a speed of 0.8 *c*?
       a. 0.6
       b. 1.2
    * c. 1.67
       d. 1.8

36. A rocket ship is 80 m long when measured at rest. What is its length as measured by an observer who sees the rocket ship moving past at 99% of the speed of light? The relativistic adjustment factor for 0.99 *c* is 7.09.
    * a. 11.3 m
       b. 79.2 m
       c. 80 m
       d. 567 m

37. Peter volunteers to serve on the first mission to visit alpha Centauri. Even traveling at 80% of the speed of light, the round trip will take a minimum of 10 years. When Peter returns from the trip, he will be _____ his twin brother Paul, who remained on Earth.
    * a. younger than
       b. the same age as
       c. older than
       d. We can't say, because the answer depends on the details of the trip.

38. The twin paradox involves the situation in which one twin remains on Earth while the other makes a trip to a distant location at a speed approaching the speed of light. When the twins are reunited on Earth, we discover that the twin who _____ is younger than the other.
       a. remained on Earth
    * b. made the trip
       c. Actually, they are the same age.
       d. We cannot say until we know the details of the trip.

39. Suppose a meter stick zips by you at a speed only slightly less than the speed of light. If you measure the length of the meter stick as it goes by, you would determine that it is _____ one meter long.
    * a. shorter than
       b. longer than
       c. still

40. A train is traveling along a straight, horizontal track at a constant speed that is only slightly less than that of light. An observer in the train determines that meter sticks on the ground are _____ meter sticks in the train.
   * a. shorter than
     b. longer than
     c. the same length as

41. A train is traveling along a straight, horizontal track at a constant speed that is only slightly less than that of light. Observers in the train and on the ground measure the distance between the rails. The value obtained by the observers on the ground is _____ that obtained by the observers in the train.
     a. shorter than
     b. longer than
   * c. the same as

42. A train is traveling along a straight, horizontal track at a constant speed that is only slightly less than that of light. An observer on the ground and an observer in the train each measure the distance between two posts located along the tracks. The observer on the ground obtains ____ the observer in the train.
   * a. a longer length than
     b. the same length as
     c. a shorter length than

43. A train is traveling along a straight, horizontal track at a constant speed that is only slightly less than that of light. An observer on the ground and an observer in the train each measure the length of the train. The observer in the train obtains ____ the observer on the ground.
   * a. a longer length than
     b. the same length as
     c. a shorter length than

44. A train is traveling along a straight, horizontal track at a constant speed that is only slightly less than that of light. An observer in the train claims that the engine came out of a tunnel at the same time as the caboose entered the tunnel. An observer on the ground claims that the engine came out of the tunnel _____ the caboose entered.
     a. at the same time as
     b. before
   * c. after

45. A train is traveling along a straight, horizontal track at a constant speed that is only slightly less than that of light. An observer on the ground claims that the engine came out of a tunnel at the same time as the caboose entered the tunnel. An observer in the train claims that the engine came out of the tunnel _____ the caboose entered.
    a. at the same time as
*   b. before
    c. after

46. An observer at rest on the ground claims that a moving long pole momentarily fits into a short barn. An observer moving with the pole claims that this is impossible since, from his reference system, the barn is short and the pole is long. The observer riding on the pole argues that the
    a. relativistic effects can only be viewed from the ground.
*   b. back door opens before the front door closes.
    c. back door opens after the front door closes.
    d. back door opens at the same time as the front door closes.

47. An electron is being accelerated by a constant force to nearly the speed of light. Which of the following is NOT true?
*   a. It experiences a constant acceleration.
    b. Its momentum increases at a constant rate.
    c. It can approach but not exceed the speed of light.
    d. Its total energy continually increases.

48. A proton is being accelerated by a constant force to nearly the speed of light. Which of the following is true?
    a. The proton experiences a constant acceleration.
*   b. The proton's momentum increases at a constant rate.
    c. The proton's momentum increases at an increasing rate.
    d. The impulse decreases with time.

49. A train is traveling along a straight, horizontal track at a constant speed that is only slightly less than that of light. Each observer measures the mass of a 1-kg mass in the train. The observer on the ground claims that its mass is _____ 1 kg.
    a. greater than
*   b. equal to
    c. less than

50. Which of the following expressions gives the total relativistic energy of an object?
    a. $E = mc^2$
*   b. $E = \gamma mc^2$
    c. $E = (\gamma - 1)mc^2$
    d. $E = 0.5\, mv^2$

Chapter 10   Einstein's Relativity

51. Which of the following does NOT agree with the expression $E = mc^2$?
    a. Energy can be converted to mass.
    b. Mass can be converted to energy.
  * c. E is the total energy of the particle.
    d. c is the speed of light.

52. The conclusions of the special theory of relativity
    a. are true only for objects moving at very high speeds.
    b. have not yet been experimentally verified.
    c. apply only to tiny atomic particles.
  * d. are believed to be true for all motions of all objects.

53. We presently believe Einstein's theory of special relativity is a correct view of nature because the predictions
    a. seem to make good common sense.
    b. can be calculated from Newton's laws.
    c. can be calculated from the conservation laws.
  * d. are in agreement with experiment.

54. Imagine a spaceship that is far from any large masses so that the effects of gravity are negligible. This spaceship has a velocity of 10,000 km/s and an acceleration in the forward direction of 10 m/s^2. If you release a ball in this ship, you will find that it falls to the floor with an acceleration _____ 10 m/s^2.
    a. less than
  * b. equal to
    c. larger than

55. The principle of equivalence in the general theory of relativity states that
    a. the laws of physics are the same in all inertial systems.
    b. all clocks are equivalent.
    c. space is warped.
  * d. acceleration and gravitation are equivalent.

56. If inertial mass and gravitational mass were NOT the same,
    a. the law of universal gravitation would need to be modified.
    b. Newton's second law would need to be modified.
  * c. all objects falling in a vacuum near Earth's surface would not experience the same acceleration.
    d. all objects falling in a vacuum near Earth's surface would no longer experience the same force.

57. The mass that appears in Newton's second law of motion is
    * a. the inertial mass.
      b. the gravitational mass.
      c. both the inertial and gravitational mass.
      d. the real mass.

58. The mass that appears in Newton's law of universal gravitation is
      a. the inertial mass.
    * b. the gravitational mass.
      c. both the inertial and gravitational mass.
      d. the real mass.

59. Two balls of different mass are simultaneously released in a vacuum in a spaceship with a constant acceleration in the forward direction. If the balls are released from the same height, which one will hit the floor first?
      a. the heavier one
      b. the lighter one
      c. the more dense one
    * d. They will both hit at the same time.

60. Suppose two teams of astronauts who think they are accelerating through space are actually sitting on the surfaces of Earth and Mercury. The gravitational field on Mercury is much smaller than that on Earth. Which team thinks it has the larger acceleration?
      a. Mercury
    * b. Earth
      c. The accelerations are the same.
      d. Neither team can determine their acceleration.

61. Suppose two teams of astronauts who think they are accelerating through space are actually sitting on the surfaces of Earth and Mercury. The gravitational field on Mercury is much smaller than that on Earth. Which team thinks it has the larger speed?
      a. Mercury
      b. Earth
      c. The speeds are the same.
    * d. Neither team can determine their speed.

62. Two spaceships are traveling through space with different accelerations. If the passengers believe that they are actually sitting on planets, which ship would be sitting on the planet with the larger acceleration due to gravity?
    * a. the one with the larger acceleration
      b. the one with the smaller acceleration
      c. The passengers cannot determine this.

63. Imagine a spaceship that is far from any large masses so that the effects of gravity are negligible. This spaceship has a velocity of 460 km/s and an acceleration in the forward direction of 6 m/s^2. What is the acceleration of a ball after it is released in this spaceship?
   * a. 6 m/s^2
      b. 10 m/s^2
      c. 60 m/s^2
      d. 460 km/s^2

64. A spaceship is resting on the surface of Mars where the acceleration due to gravity is 40% of that on Earth. If the astronauts think that they are accelerating through space, what would their acceleration be?
   * a. 4 m/s^2
      b. 8 m/s^2
      c. 10 m/s^2
      d. 40 m/s^2

65. A special train is traveling down a long, straight track with a constant acceleration of 10 m/s^2 in the forward direction. If a passenger drops a ball and watches its path, the ball falls
      a. straight down.
      b. slightly toward the back of the train.
   * c. backward at a 45° angle.
      d. forward at a 45° angle.

66. A special train is traveling down a long, straight track with a constant acceleration of 3 m/s^2 in the forward direction. The train is located on Mars where the acceleration due to gravity is 4 m/s^2. If a passenger drops a ball and measures the acceleration of the ball, the acceleration will be
      a. 3 m/s^2
      b. 4 m/s^2
   * c. 5 m/s^2
      d. 10 m/s^2

67. The general theory of relativity says that light is _____ a large mass.
      a. not affected by
   * b. attracted toward
      c. repelled from

68. A "black hole" derives its name from the observation that
      a. it is a void in space.
   * b. light cannot escape its gravitational field.
      c. the light from its surface resembles that from a "black" light.
      d. X rays are not visible to the human eye.

69. The general theory of relativity says that clocks run _____ as the gravitational force is increased.
    a. at the same rate
  * b. slower
    c. faster

70. Would clocks on board a space shuttle gain or lose time during a launch?
    a. gain
  * b. lose
    c. The clocks would not be affected.

71. The Einsteinian world view says that
    a. objects travel in straight lines in three-dimensional space.
    b. the presence of matter introduces forces that causes objects to move along curved paths.
  * c. the presence of matter warps space-time.
    d. space obeys the rules of Euclidean geometry.

72. The conclusions of the general theory of relativity
    a. are true only for objects moving at very high speeds.
    b. have not yet been experimentally verified.
    c. apply only to tiny atomic particles.
  * d. are believed to be true for all motions of all objects.

73. We presently believe Einstein's theory of general relativity is a correct view of nature because the predictions
    a. seem to make good common sense.
    b. can be calculated from Newton's laws.
    c. can be calculated from the conservation laws.
  * d. are in agreement with experiment.

Chapter 11: Structure of Matter

1. Suppose that you are given a sealed can containing a liquid. Which of the following conclusions could you NOT make without opening the can?
   - a. The liquid freezes at 5°C.
   - b. The mass of the liquid is less than 246 grams.
   - * c. The liquid has a sweet smell.
   - d. The consistency of the liquid is more like molasses than water.

2. Which of the following is NOT necessarily a feature of a good model?
   - a. It must account for the known data.
   - * b. It must be expressible in mathematical form.
   - c. It must agree with the way nature behaves.
   - d. It must be able to make predictions about new situations.

3. Which of the following is NOT one of the Aristotelian elements?
   - a. Earth
   - * b. oxygen
   - c. fire
   - d. water

4. The Greek "atomists" believed in atoms
   - a. because of experiments with combining gases.
   - b. because of diffusion experiments.
   - * c. on philosophic arguments.
   - d. because they believed the alchemists.

5. Which of the following is a compound?
   - a. hydrogen
   - b. oxygen
   - c. carbon
   - * d. water

6. Which two elements combine to form table salt?
   - a. hydrogen and oxygen
   - b. sodium and oxygen
   - * c. sodium and chlorine
   - d. hydrogen and chlorine

7. Water and sugar form a
   - * a. mixture.
   - b. element.
   - c. compound.
   - d. molecule.

8. Sodium and chlorine form a
   a. mixture.
   b. element.
   * c. compound.
   d. dichotomy.

9. The law of definite proportions states that _____ have definite _____ ratios of their constituent elements.
   * a. compounds ... mass
   b. compounds ... volume
   c. mixtures ... mass
   d. mixtures ... volume

10. If 1 gram of hydrogen combines completely with 8 grams of oxygen to form water, how many grams of hydrogen does it take to combine completely with 32 grams of oxygen?
    a. 2
    b. 3
    * c. 4
    d. 5

11. If 3 grams of carbon combine completely with 4 grams of oxygen to form carbon monoxide, how many grams of carbon does it take to combine completely with 20 grams of oxygen?
    a. 7
    b. 12
    * c. 15
    d. 20

12. If 4 grams of oxygen combine completely with 3 grams of carbon to form carbon monoxide (1 carbon atom and 1 oxygen atom in each molecule), how many grams of oxygen does it take to combine completely with 3 grams of carbon to form carbon dioxide (1 carbon atom and 2 oxygen atoms)?
    a. 2
    b. 4
    c. 6
    * d. 8

13. In ammonia, 14 g of nitrogen combines completely with 3 g of hydrogen. How many grams of hydrogen does it take to combine completely with 56 g of nitrogen?
    a. 3 g
    b. 4 g
    * c. 12 g
    d. 56 g

14. Given that 1 g of hydrogen combines completely with 8 g of oxygen to form water, how many grams of water can you make with 8 g of hydrogen and 32 g of oxygen?
    a. 4 g
    b. 32 g
    * c. 36 g
    d. 40 g

15. Given that 12 g of carbon combines completely with 16 g of oxygen to form carbon monoxide, how many grams of carbon monoxide can be made from 36 g of carbon and 90 g of oxygen?
    a. 48 g
    * b. 84 g
    c. 90 g
    d. 126 g

16. A ham sandwich consists of one slice of ham (10 g) and two slices of bread (25 g each). You have 1 kg of ham and 1 kg of bread. You make as many sandwiches as you can. How many sandwiches did you make?
    * a. 20
    b. 40
    c. 100
    d. 120

17. A ham sandwich consists of one slice of ham (10 g) and two slices of bread (25 g each). You have 1 kg of ham and 1 kg of bread. You make as many sandwiches as you can. What is the mass of the sandwiches?
    a. 1.0 kg
    * b. 1.2 kg
    c. 1.4 kg
    d. 2.0 kg

18. One mole of water molecules consists of 1 mole of oxygen (16 g) and 2 moles of hydrogen (1 g each). You combine 1 kg of oxygen with 1 kg of hydrogen to make water. How many moles of water did you make?
    a. 3
    * b. 62.5
    c. 562.5
    d. 1062.5

19. One mole of water molecules consists of 1 mole of oxygen (16 g) and 2 moles of hydrogen (1 g each). You combine 1 kg of oxygen with 1 kg of hydrogen to make water. What is the mass of the resulting water?
    a. 1000 g
    b. 1062.5 g
  * c. 1125 g
    d. 2000 g

20. Avogadro suggested that each liter of gas under identical conditions has the same
    a. mass.
    b. number of atoms.
  * c. number of molecules.
    d. density.

21. How does the number of molecules in 1 liter of oxygen compare with the number of molecules in 1 liter of carbon dioxide if they are both at the same temperature and pressure?
  * a. They have the same number of molecules.
    b. Oxygen has more molecules
    c. Carbon dioxide has more molecules.

22. Assume that you have equal volumes of oxygen and hydrogen at the same temperature and pressure. If each molecule of oxygen and hydrogen contains two atoms, how do the numbers of oxygen and hydrogen atoms in the gases compare?
  * a. They are the same.
    b. The oxygen has more.
    c. The hydrogen has more.

23. If 1 liter of oxygen combines completely with 2 liters of hydrogen to form water, how many liters of oxygen are required to combine completely with 12 liters of hydrogen?
    a. 4
  * b. 6
    c. 10
    d. 12

24. One liter of nitrogen (N) combines with 3 liters of hydrogen (H) to form 2 liters of ammonia. If the molecules of nitrogen and hydrogen have 2 atoms each, how many atoms of hydrogen and nitrogen are there in 1 molecule of ammonia?
    a. 2 H and 1 N
    b. 2 H and 2 N
  * c. 3 H and 1 N
    d. 3 H and 2 N

Chapter 11   Structure of Matter

25. Three grams of carbon combine completely with 4 grams of oxygen to form carbon monoxide. The carbon monoxide molecule consists of one atom of each kind and the oxygen atom has an atomic mass of 16. What is the atomic mass of carbon?
    a. 3
    b. 4
*   c. 12
    d. 16

26. If an atom of titanium is four times as massive as an atom of carbon, what is the atomic mass of titanium?
    a. 3 amu
    b. 12 amu
    c. 24 amu
*   d. 48 amu

27. The atomic mass of carbon monoxide is 28 atomic mass units. What is the atomic mass of oxygen if each molecule contains one atom of carbon and one of oxygen?
    a. 12
    b. 14
*   c. 16
    d. 22

28. The atomic mass of carbon dioxide is 44 atomic mass units. What is the atomic mass of oxygen if each molecule contains one atom of carbon and two of oxygen?
    a. 10
    b. 12
*   c. 16
    d. 22

29. If the atomic mass of neon is 20 amu, how much neon would be needed to have an Avogadro's number of neon atoms?
    a. 20 amu
    b. 10 grams
*   c. 20 grams
    d. 20 kilograms

30. Given that the carbon atom has a mass of 12 amu, how many carbon atoms are there in a diamond with a mass of 1 g?
    a. 6.02×10^{22}
*   b. 5.02×10^{22}
    c. 1.00×10^{23}
    d. 6.02×10^{23}

31. Given that the sulfur molecule has a mass of 32 amu, how many sulfur molecules are in 1 g of sulfur?
    a. 6.02×10^{22}
  * b. 1.88×10^{22}
    c. 6.02×10^{23}
    d. 1.93×10^{25}

32. One liter of water has a mass of 1 kg, and the mass of a water molecule is 18 amu. How many molecules of water are there in 1 L of water?
    a. 55.5
    b. 1.08×10^{22}
    c. 6.02×10^{23}
  * d. 3.34×10^{25}

33. One liter of oxygen has a mass of 1.4 g, and the oxygen molecule has a mass of 32 amu. How many oxygen molecules are there in 1 L of oxygen?
    a. 44.8
  * b. 2.63×10^{22}
    c. 6.02×10^{23}
    d. 2.70×10^{25}

34. Which of the following is NOT a feature of our ideal gas? The gas particles
  * a. are massless.
    b. have no internal structure.
    c. are indestructible.
    d. do not interact except when they collide.

35. The pressure that a gas exerts on the walls of its container is a direct result of the
    a. repulsive forces between gas molecules.
    b. combined volume of the gas molecules.
  * c. collisions of the gas molecules with the walls.
    d. combined mass of the gas molecules.

36. If the heel of a woman's shoe has an area of 1 square centimeter and the woman has a mass of 70 kilograms, what pressure can she exert on the floor if she puts all of her weight on one heel?
    a. 70 kg·cm^2
    b. 70 kg/cm^2
    c. 70 N/cm^2
  * d. 700 N/cm^2

Chapter 11   Structure of Matter

37. You exert a force of 30 N on the head of a thumbtack. The head of the thumbtack has a radius of 5 mm. What is the pressure on your thumb?
    a. 2.62×10^{-6} Pa
    b. 0.382 Pa
    c. 2.62 Pa
*   d. 3.82×10^{5} Pa

38. If a liter of gas has a pressure of 1 atmosphere, what will the pressure be if the average kinetic energy of the molecules is doubled?
    a. 0.5 atm
    b. 1 atm
*   c. 2 atm
    d. 4 atm

39. If the column in an alcohol-in-glass thermometer were not uniform, the spacing between degrees on a narrow portion of the thermometer would be _____ those on a wide portion.
    a. closer than
    b. the same as
*   c. wider than

40. The two fixed points used to define the modern Fahrenheit temperature scale are those of
    a. boiling water and a mixture of ice and salt.
    b. the body and a mixture of ice and salt.
    c. the body and freezing water.
*   d. boiling water and freezing water.

41. Why is body temperature NOT a good fixed temperature for establishing a temperature scale?
    a. It is not hot enough.
    b. It is not a whole number.
*   c. It varies for an individual and among individuals.
    d. Body temperature cannot be measured until a temperature scale is developed.

42. Two students are sick in bed with fevers. Bill has a temperature of 2°C above normal, while Sally's is 2°F above normal. Which student has the higher fever?
*   a. Bill
    b. Sally
    c. They have the same fevers.

43. Is a sauna at a temperature of 202°F hotter or colder than one at 90°C?
*   a. hotter
    b. colder
    c. They have the same temperature.

44. What Fahrenheit temperature corresponds most closely to 45°C?
    a. 49°F
    b. 57°F
    c. 99°F
    * d. 113°F

45. What Celsius temperature corresponds most closely to 77°F?
    * a. 25°C
    b. 43°C
    c. 60°C
    d. 81°C

46. A student decides to devise a new temperature scale with the freezing and boiling points at 0°X and 70°X. What Celsius temperature would correspond to a temperature of 35°X?
    a. 35
    * b. 50
    c. 70
    d. 90

47. The Kelvin temperature scale has the same size degree as the_____ scale and its zero is _____ degrees lower.
    a. Fahrenheit ... 459
    b. Fahrenheit ... 273
    c. Celsius ... 459
    * d. Celsius ... 273

48. Which of the following statements is true for an ideal gas? The average _____ of an ideal gas is proportional to the _____ temperature.
    a. speed ... Celsius
    b. kinetic energy ... Celsius
    c. speed ... Kelvin
    * d. kinetic energy ... Kelvin

49. Which of the following doubles with a doubling of the absolute temperature of an ideal gas?
    a. average momentum
    b. average speed
    c. average velocity
    * d. average kinetic energy

50. Which of the following doubles with a doubling of the Celsius temperature of an ideal gas?
         a. average momentum
         b. average speed
         c. average kinetic energy
    *   d. None of the above

51. By what factor does the absolute temperature of an ideal gas increase when the root mean square speed is doubled?
         a. 1
         b. 2
    *   c. 4
         d. 8

52. What Kelvin temperature corresponds to room temperature of $20^\circ C$?
         a. 20 K
         b. 253 K
         c. 273 K
    *   d. 293 K

53. The pressure in a container filled with gas increases when it is heated because
         a. the walls do work on the gas.
    *   b. the average momentum of each gas particle increases in size.
         c. the number of gas particles increases.
         d. the volume of the gas decreases.

54. Two gases are kept at the same temperature. If the molecules of gas A have 4 times the mass of those of gas B, what is the ratio of the rms speed of the A molecules to that of the B molecules?
         a. 4
         b. 2
    *   c. 1/2
         d. 1/4

55. The mass of an oxygen molecule is 16 times that of a hydrogen molecule. If the gases are maintained at the same temperature, what is the ratio of the rms speed of an oxygen molecule to that of a hydrogen molecule?
         a. 1/16
    *   b. 1/4
         c. 4
         d. 16

56. What happens to the temperature of an ideal gas if the volume is reduced to one-half while holding the pressure constant? The temperature
    a. quadruples.
    b. doubles.
*   c. is cut in half.
    d. is cut to one-fourth.

57. If you hold the temperature of an ideal gas constant, what happens to its volume if you double its pressure? The volume
    a. quadruples.
    b. doubles.
*   c. is cut in half.
    d. is cut to one-fourth.

58. If the volume of an ideal gas is held constant, what happens to the temperature if the pressure is doubled? The temperature
    a. quadruples.
*   b. doubles.
    c. is cut in half.
    d. is cut to one-fourth.

59. Four liters of an ideal gas are heated from 300 K to 600 K in a container with a fixed volume? If the initial pressure was 1 atm, what is the final pressure?
    a. 4 atm
*   b. 2 atm
    c. 0.5 atm
    d. 0.25 atm

60. Two liters of an ideal gas are heated from 300 K to 600 K while the pressure is maintained at 1 atm. What is the final volume of the gas?
    a. 16 liters
    b. 8 liters
*   c. 4 liters
    d. 2 liters

Chapter 12: States of Matter

1. Which of the four states of matter occurs at the highest temperature?
   * a. plasma
   b. liquid
   c. solid
   d. gas

2. What is the force that binds atoms together?
   a. gravitational
   * b. electric
   c. magnetic
   d. nuclear

3. What is the force that binds materials together?
   a. gravitational
   * b. electric
   c. strong
   d. weak

4. What is the distinguishing characteristic of crystals?
   a. They are solids.
   * b. They have repeating geometrical structures.
   c. They are bonded together by the electric forces between atoms.
   d. They have low melting points.

5. What of the following is NOT a crystal?
   a. mica
   b. ice
   * c. bronze
   d. table salt

6. Window glass is an example of
   a. a crystal.
   * b. an amorphous solid.
   c. an alloy.
   d. a polymer.

7. Density is defined as
   a. weight per unit volume.
   b. weight per unit area.
   * c. mass per unit volume.
   d. mass per unit area.

8. Which has the larger density, a gold brick or a gold coin?
     a. brick
     b. coin
 *   c. They have the same density.

9. A cubic meter of metal A has a mass 30% larger than a cubic meter of metal B. Which metal has the smaller density?
     a. metal A
 *   b. metal B
     c. They have the same density.

10. A kilogram of metal A has a volume 20% larger than a kilogram of metal B. Which metal has the smaller density?
 *   a. metal A
     b. metal B
     c. They have the same density.

11. Which has the larger density, fatty tissue or muscle?
     a. fatty tissue
 *   b. muscle
     c. They have the same density.

12. If a block of material has a mass of 16 grams and a volume of 8 cubic centimeters, what is its density?
     a. 0. 5 cubic centimeters per gram
 *   b. 2 grams per cubic centimeter
     c. 24 grams per cubic centimeter
     d. 128 grams-cubic centimeters

13. If a plastic toy has a mass of 9 g and a volume of 3 cm^3, what is its density?
     a. 0.33 g/cm^3
 *   b. 3 g/cm^3
     c. 6 g/cm^3
     d. 12 g/cm^3

14. Iron has a density of 7860 kg/m^3. What is the density of copper in g/cm^3?
 *   a. 7.86
     b. 78.6
     c. 786
     d. 7860

Chapter 12   States of Matter

15. Uranium has a density of 18.7 g/cm^3. What is the density of silver in kg/m^3?
    a. 18.7
    b. 187
    c. 1870
* d. 18,700

16. What is the volume of a metallic object with a mass of 36 g if its density is 12 g/cm^3?
    a. 0.33 m^3
* b. 3 cm^3
    c. 24 cm^3
    d. 48 cm^3

17. If the density of a certain kind of wood is 0.8 g/cm^3, what is the mass of 5 cm^3 of the wood?
    a. 0.4 g
* b. 4.0 g
    c. 4.2 g
    d. 5.8 g

18. Given that most people are just about neutrally buoyant, it is reasonable to estimate the density of the human body to be about that of water. Given this assumption, the volume of an 80-kg person is
* a. 0.08 m^3
    b. 0.8 m^3
    c. 8 m^3
    d. 80 m^3

19. A cube with a mass of 54 g is made from a metal with a density of 6 g/cm^3. The volume of the cube is
* a. 9 cm^3
    b. 48 cm^3
    c. 60 cm^3
    d. 324 cm^3

20. If 1000 cm^3 of a gas with a density of 0.0009 g/cm^3 condenses to a liquid with a density of 0.9 g/cm^3, what is the volume of the liquid?
* a. 1 cm^3
    b. 10 cm^3
    c. 100 cm^3
    d. 1000 cm^3

21. A solid ball with a volume of 0.4 m^3 is made of a material with a density of 2500 kg/m^3. What is the mass of the ball?
    * a. 1000 kg
    b. 1500 kg
    c. 2500 kg
    d. 6250 kg

22. A spring which obeys Hooke's law stretches 3 cm when 12 g is hung from it. How much does the spring stretch when 48 g is hung from it?
    a. 4 cm
    b. 6 cm
    c. 9 cm
    * d. 12 cm

23. When a 400-g mass is hung from the end of a spring, the spring stretches 6 cm. How much will the spring stretch when a 100-g mass is used?
    * a. 1.5 cm
    b. 6 cm
    c. 24 cm
    d. 50 cm

24. What is the spring constant for a spring which stretches 4 cm when we hang 8 kg from it?
    a. 2 N/m
    b. 20 N/m
    c. 200 N/m
    * d. 2000 N/m

25. How do liquid crystals differ from ordinary liquids? Liquid crystals have
    a. higher melting points.
    b. repeating geometrical patterns.
    c. higher densities.
    * d. an orientational order.

26. Which of the following effects is NOT a result of surface tension?
    a. A glass can be filled with milk beyond its brim.
    b. Liquids floating in the Space Shuttle from into spheres.
    * c. The pressure increase with the depth in a fluid.
    d. Soap films on wire frames have minimal surface areas.

27. Which of the following has the largest viscosity?
    * a. honey
      b. air
      c. water
      d. gasoline

28. Pressure is the
      a. mass per unit volume.
      b. mass per unit area.
      c. force per unit volume.
    * d. force per unit area.

29. In which direction in a fluid at rest can you move without the pressure changing?
      a. up
      b. down
    * c. horizontal
      d. The pressure doesn't change as you move in any of these directions.

30. In which direction in a fluid at rest is the pressure the greatest at a given depth?
      a. upward
      b. downward
      c. horizontal
    * d. The pressure is the same in all directions.

31. Which of the following is NOT a unit used for atmospheric pressure?
      a. millimeters of mercury
      b. pounds per square inch
    * c. newtons
      d. pascals

32. Why don't the forces exerted inward on our bodies by the atmosphere crush our bodies?
      a. The forces are too small due to the low density of air.
      b. The outer surfaces of the cells are strong enough to withstand the forces.
    * c. The pressure inside our bodies is essentially the same as outside.
      d. Evolution has made our bodies capable of withstanding these strong forces.

33. At an atmospheric pressure of 100 kilopascals, what is the weight of the column of air above a square meter of the earth's surface?
      a. 100 N
      b. 147 N
    * c. 100,000 N
      d. 147,000 N

34. Two barometers are made with water and oil. The density of oil is 900 kg/m^3 and that of water is 1000 kg/m^3. If the oil column is 10 m tall, how tall is the water column?
    a. 11.1 m
    b. 10 m
    * c. 9 m
    d. 8.1 m

35. On a day when the mercury column in a barometer on top of a mountain stands 50 cm tall, how high would the column in a water barometer be? The density of mercury is 13.6 times that of water.
    a. 3.68 cm
    b. 50 cm
    c. 63.6 cm
    * d. 680 cm

36. On a day when a water barometer reads 10 m, what is the pressure at the bottom of a 20-m deep tank of water?
    a. 1 atm
    b. 2 atm
    * c. 3 atm
    d. 4 atm

37. Archimedes' principle tell us that the buoyant force is equal to the
    a. mass of the object.
    b. weight of the object.
    c. mass of the displaced fluid.
    * d. weight of the displaced fluid.

38. An object floats whenever its
    a. mass is less than the mass of an equal volume of the fluid.
    b. weight is less than the weight of an equal volume of the fluid.
    c. density is less than the density of the fluid.
    * d. All of the above.

39. If you weigh a block of metal under water, its weight is _____ its weight in air.
    * a. less than
    b. larger than
    c. the same as

40. Salt water is slightly denser than fresh water. A 50-ton ship will experience a greater buoyant force floating in a lake of
    a. fresh water.
    b. salt water.
    * c. The buoyant force would be the same in both lakes.

41. Salt water is slightly denser than fresh water. Will a 12-pound bowling ball feel a greater buoyant force sitting on the bottom of a freshwater lake or on the bottom of the ocean?
    a. on the bottom of a freshwater lake
    * b. on the bottom of the ocean
    c. The buoyant force would be the same.

42. You have two cubes of the same size, one made of aluminum and the other of lead. Both cubes are allowed to sink to the bottom of a water-filled aquarium. Which cube, if either, experiences the greater buoyant force?
    a. the aluminum cube
    b. the lead cube
    * c. Both cubes experience the same buoyant force.

43. You have two cubes of the same size, one made of wood and the other of aluminum. Both cubes are placed in a water-filled aquarium. The wooden block floats, and the aluminum block sinks. Which cube, if either, experiences the greater buoyant force?
    a. the wood cube
    * b. the aluminum cube
    c. Both cubes experience the same buoyant force.

44. An ice cube is floating in a glass of water. As the ice cube melts, the water level in the glass will
    a. go up.
    b. go down.
    * c. stay the same.

45. Three spheres have the same mass and all float in water. Spheres A, B, and C have volumes of 30 cm^3, 35 cm^3, and 40 cm^3, respectively. Which sphere floats the highest?
    a. A
    b. B
    * c. C
    d. They all float at the same height.

46. Three spheres have the same volumes and all float in water. Spheres A, B, and C have masses of 25 g, 30 g, and 35 g, respectively. Which sphere floats the highest?
    * a. A
    b. B
    c. C
    d. They all float at the same height.

47. A block of plastic has a mass of 50 g and a volume of 40 cm^3. Will the block sink or float in water?
    * a. sink to the bottom
    b. float with 20% of the block above the surface
    c. float with its top at the surface
    d. sink to 80% of the depth of the water

48. A 500-g wooden block is lowered carefully into a completely full beaker of water and floats. What is the weight of the water, in newtons, that spills out of the beaker?
    * a. 5 N
    b. 50 N
    c. 500 N
    d. 5000 N

49. A 400-cm^3 block of aluminum ($D = 2.7$ g/cm^3) is lowered carefully into a completely full beaker of water. What is the weight of the water that spills out of the beaker?
    a. 1.08 N
    b. 1.48 N
    * c. 4 N
    d. 10.8 N

    Difficulty: 2

50. A piece of plastic just barely floats in water. If it has a mass of 20 g, what is its volume?
    a. 2 cm^3
    b. 10 cm^3
    * c. 20 cm^3
    d. 30 cm^3

51. A hollow box has a mass of 200 g. What is the minimum volume the box could have and still float?
    a. 20 cm^3
    b. 100 cm^3
    * c. 200 cm^3
    d. 300 cm^3

52. A plastic bobber just barely floats in water. If it weighs 2 N in air, what is its volume?
    a. 2 cm^3
    b. 100 cm^3
    * c. 200 cm^3
    d. 300 cm^3

53. A hollow sphere has an average density of 3 g/cm^3 and a mass of 120 g. What will the sphere weigh under water?
    * a. 0.8 N
    b. 0.9 N
    c. 0.12 N
    d. 0.36 N

54. A block of metal weighs 9 N in air and 7 N in water. What is the volume of the block?
    a. 2 cm^3
    * b. 200 cm^3
    c. 700 cm^3
    d. 900 cm^3

55. An incompressible fluid is being pumped through a pipe that has a narrow region. The pressure in the narrow region is _____ that in the wider regions of the pipe.
    * a. less than
    b. the same as
    c. greater than

56. Which of the following is NOT due to the Bernoulli effect?
    a. raised tarpaulins on moving trucks
    b. the curve of a baseball
    c. the lift produced on airplane wings
    * d. the floating of icebergs

Chapter 13: Thermal Energy

1. Heat is the
   a. same as temperature.
   * b. thermal energy that is transferred from one object to another.
   c. potential energy associated with temperature.
   d. massless fluid generated by doing work on the system.

2. Count Rumford's experiments with boring cannons
   * a. showed that heat was not a fluid.
   b. showed that 4.2 joules of work are equivalent to 1 calorie of heat.
   c. were used to define the calorie.
   d. confirmed the law of conservation of energy.

3. A calorie is defined to be the amount of heat required to raise the temperature of
   a. 1 pound of water by $1^{\circ}C$.
   * b. 1 gram of water by $1^{\circ}C$.
   c. 1 gram of water by $1^{\circ}F$.
   d. 1 pound of water by $1^{\circ}F$.

4. A Btu is defined to be the amount of heat required to raise the temperature of
   a. 1 pound of water by $1^{\circ}C$.
   b. 1 gram of water by $1^{\circ}C$.
   c. 1 gram of water by $1^{\circ}F$.
   * d. 1 pound of water by $1^{\circ}F$.

5. How many calories are required to heat 600 g of water from $23^{\circ}C$ to $33^{\circ}C$?
   a. 60
   b. 600
   * c. 6000
   d. 13,800

6. How many Btu's are required to heat 6 lbs of water from $40^{\circ}F$ to $50^{\circ}F$?
   a. 6
   * b. 60
   c. 240
   d. 300

7. How much heat is required to raise the temperature of 500 g of water from $20^{\circ}C$ to $50^{\circ}C$?
   a. 500 cal
   b. 10,000 cal
   * c. 15,000 cal
   d. 25,000 cal

Chapter 13   Thermal Energy

8. If the temperature of 500 g of water drops by 8°C, how much heat is released?
   - a. 62.5 cal
   - b. 500 cal
   - c. 2000 cal
   - * d. 4000 cal

9. If 200 g of water at 100°C are mixed with 300 g of water at 50°C in a completely insulated container, what is the final equilibrium temperature?
   - a. 50 °C
   - * b. 70 °C
   - c. 80 °C
   - d. 100 °C

10. What effect would there be on Joule's experiment if the mass fell rapidly, hit the floor with substantial speed, and this was ignored?
    - a. Nothing. The loss in gravitational potential energy would still be the same.
    - b. Nothing. The kinetic energy is converted to heat when the mass hits the floor.
    - * c. The number of joules equivalent to 1 calorie would be larger.
    - d. The number of joules equivalent to 1 calorie would be smaller.

11. Joule's experiments with hanging weights turning paddle wheels in water
    - a. showed that heat was not a fluid.
    - * b. showed that 4.2 joules of work are equivalent to 1 calorie of heat.
    - c. were used to define the calorie.
    - d. showed that heat could be converted 100% to mechanical energy.

12. What would you expect the temperature of water at the bottom of Niagara Falls to be relative to the water temperature at the top? For simplicity, assume that there is no heat loss to the air.
    - a. the same
    - b. hotter at the bottom
    - * c. cooler at the bottom

13. How many joules are equivalent to 21 calories?
    - a. 5
    - b. 17
    - c. 25
    - * d. 88

Physics: A World View, Sixth Edition by Larry Kirkpatrick and Gregory Francis

14. How many calories are equivalent to 21 joules?
    * a. 5
    b. 17
    c. 25
    d. 88

15. A typical jogger burns up food energy at the rate of about 40 kJ per minute. How long would it take to run off a piece of cake if it contains 400 Calories (about 1,700 kJ)?
    a. 1 min
    b. 4.25 min
    c. 10 min
    * d. 42.5 min

16. Which of the following statements does NOT correctly describe what happens when a hot block is placed in contact with a cool block?
    a. Heat flows from the hot block to the cool block.
    b. The average kinetic energy of the particles decreases in the hot block and increases in the cool block.
    c. The temperature of the hot block decreases and that of the cool block increases.
    * d. Temperature flows from the hot block to the cool block.

17. Two objects are in thermal equilibrium if
    a. they have the same temperature.
    b. they are each in thermal equilibrium with a third object.
    c. they are in thermal contact and there is no net flow of thermal energy.
    * d. All of the above are true.

18. The zeroth law of thermodynamics
    a. is a restatement of the law of conservation of energy.
    b. says that heat cannot be completely converted to mechanical energy.
    * c. is the basis for the definition of temperature.
    d. is the basis for the definition of internal energy.

19. Which law of thermodynamics is the basis for the definition of temperature?
    * a. zeroth
    b. first
    c. second
    d. third

123

Chapter 13   Thermal Energy

20. Which law of thermodynamics is the basis for the definition of internal energy?
        a. zeroth
    *   b. first
        c. second
        d. third

21. The first law of thermodynamics
    *   a. is a restatement of the law of conservation of energy.
        b. says that heat cannot be completely converted to mechanical energy.
        c. is the basis for the definition of temperature.
        d. is the basis for the definition of entropy.

22. The first law of thermodynamics
        a. is the basis for the definition of entropy.
        b. is the basis for the definition of temperature.
    *   c. is the basis for the definition of internal energy.
        d. says that heat cannot be completely converted to mechanical energy.

23. The first law of thermodynamics is a re-statement of the law of conservation of
        a. angular momentum.
        b. linear momentum.
        c. mechanical energy.
    *   d. energy.

24. The first law of thermodynamics is valid only
        a. if no work is done on the system.
        b. when there is no friction.
        c. if there is no heat loss or gain.
    *   d. if the system is isolated.

25. If a system has no change in internal energy, we can say that
        a. the system lost no heat.
        b. no work was done on the system.
    *   c. the amount of work done by the system was equal to the heat gained.
        d. the change in heat energy produced a temperature change.

26. Which of the following does NOT determine the amount of internal energy an object has?
        a. temperature
        b. amount of material
        c. type of material
    *   d. shape of the object

27. Which of the following statements about a cup of water and a gallon of water at the same temperature is correct?
    - a. They can transfer the same heat energy.
    - b. They have the same internal energies.
    - * c. Their internal energies are proportional to their masses.
    - d. The average molecular speed in the cup of water is less.

28. During a process, 10 joules of heat are transferred into a system, while the system itself does 4 joules of work. The internal energy of the system
    - * a. increases by 6 joules.
    - b. decreases by 4 joules.
    - c. increases by 14 joules.
    - d. remains the same.

29. During a process, 10 joules of work are performed on a system, while the system gives off 4 joules of heat. The internal energy of the system
    - * a. increases by 6 joules.
    - b. decreases by 4 joules.
    - c. increases by 14 joules.
    - d. remains the same.

30. If the internal energy of an ideal gas increases by 70 J when 150 J of work are done to compress it, how much heat is released?
    - a. 70 J
    - * b. 80 J
    - c. 150 J
    - d. 230 J

31. When an ideal gas was compressed, its internal energy increased by 90 J and it gave off 60 J of heat. How much work was done on the gas?
    - a. 30 J
    - b. 60 J
    - c. 90 J
    - * d. 150 J

32. During a process, 28 J of heat are transferred into a system, while the system itself does 42 J of work. What is the change in the internal energy of the system?
    - a. −70 J
    - * b. −14 J
    - c. +14 J
    - d. +70 J

33. If the internal energy of an ideal gas increases by 150 J at the same time that the gas expands and does 240 J of work on its surroundings, how much heat has been added to the gas?
>   a. 90 J
>   b. 150 J
>   c. 240 J
> * d. 390 J

34. The third law of thermodynamics
>   a. is a restatement of the law of conservation of energy.
>   b. says that heat cannot be completely converted to mechanical energy.
> * c. says that we can never reach the absolute zero of temperature.
>   d. says that all motion ceases at absolute zero.

35. Why do climates near the coasts tend to be more moderate than near the middle of the continent?
> * a. Because water has a relatively high specific heat.
>   b. Because water has a high latent heat of vaporization.
>   c. Because the coasts have lower elevations.
>   d. Because it rains a lot on the coasts.

36. Why does the coldest part of winter occur during late January and February when the shortest day of the year is near December 21?
>   a. The specific heat of the ground is a lot higher than that of water.
>   b. The Sun is farthest from Earth in February.
>   c. Because that's when the Southern Hemisphere has summer.
> * d. It takes the release of a lot of thermal energy for the ground to cool off.

37. Given that ice has a specific heat that is one-half that of water, does it take more thermal energy to raise the temperature of 5 grams of water or 5 grams of ice by 6°C?
> * a. water
>   b. ice
>   c. It takes the same thermal energy for each one.

38. Given that ice has a specific heat that is one-half that of water, does it take more thermal energy to raise the temperature of 6 grams of water by 8°C or 5 grams of ice by 20°C?
> * a. water
>   b. ice
>   c. It takes the same thermal energy for each one.

39. Aluminum and air have almost the same specific heats. Therefore, 100 calories of heat will raise the temperature of 1 liter of aluminum _____ 1 liter of air.
    a. more than
    * b. less than
    c. the same as

40. How much energy does it take to raise the temperature of 40 g of aluminum from $60^{\circ}C$ to $90^{\circ}C$? The specific heat of aluminum is 0.2 cal/g·$^{\circ}$C.
    a. 80 cal
    b. 120 cal
    * c. 240 cal
    d. 1200 cal

41. How much energy does it take to raise the temperature of 100 g of aluminum from $50^{\circ}C$ to $80^{\circ}C$? The specific heat of aluminum is 900 J/kg·K.
    a. 90 J
    * b. 2700 J
    c. 13,500 J
    d. 2,700,000 J

42. It takes 250 cal to raise the temperature of a metallic ring from 20°C to 30°C. If the ring has a mass of 90 g, what is the specific heat of the metal?
    a. 0.14 cal/g·$^{\circ}$C
    * b. 0.28 cal/g·$^{\circ}$C
    c. 1.4 cal/g·$^{\circ}$C
    d. 2.8 cal/g·$^{\circ}$C

43. If it takes 3400 cal to raise the temperature of a 500-g statue by 44°C, what is the specific heat of the material used to make the statue?
    * a. 0.15 cal/g·$^{\circ}$C
    b. 6.8 cal/g·$^{\circ}$C
    c. 77 cal/g·$^{\circ}$C
    d. 150 cal/g·$^{\circ}$C

44. The boiling point of liquid nitrogen at atmospheric pressure is 77 K. Which of the following temperatures is the closest to the temperature of an open container of liquid nitrogen?
    a. 76 K
    * b. 77 K
    c. 78 K
    d. 293 K

45. What would happen to a pot of water on a hot stove if there were no latent heat of vaporization required for converting water to steam?
      a. The water would not boil.
      b. The water would boil at a higher temperature.
*   c. The water would all turn to steam very rapidly.
      d. The water would not form steam.

46. Why is steam at 100°C more dangerous than water at 100°C?
      a. The steam is hotter.
*   b. The steam has more internal energy per gram.
      c. The steam has a higher specific heat.
      d. The steam has less viscosity.

47. On nights when freezing temperatures are expected, why do some fruit growers turn on sprinkler systems in their orchards?
      a. The heat in the water keeps the fruit from freezing.
      b. Trees don't freeze when they are growing.
*   c. The heat released as the water freezes keeps the fruit from freezing.
      d. The water reduces the amount of oxygen near the fruit.

48. The latent heat of melting for water is 334 kJ/kg. How much energy would it take to melt 3 kg of ice at 0°C to form water at 0°C?
      a. 111 kJ
      b. 334 kJ
      c. 668 kJ
*   d. 1000 kJ

49. The latent heat of melting for water is 334 kJ/kg. How much energy would it take to melt 1 g of ice at 0°C to form water at 0°C?
      a. 14 cal
*   b. 80 cal
      c. 14,000 cal
      d. 80,000 cal

50. Which of the following is NOT a method of transporting thermal energy from one place to another?
      a. radiation
*   b. condensation
      c. conduction
      d. convection

51. Which of the following does NOT determine the rate at which heat is conducted through a slab of material?
    a. type of material
    b. difference in temperature on the two sides
    c. thickness of the slab
    * d. average temperature of the slab

52. Which of the following is the best insulating material?
    * a. static air
    b. glass
    c. wood
    d. brick

53. Which of the following is the best thermal conductor?
    a. ceramic
    b. water
    c. wood
    * d. copper

54. Which type of bench would feel the warmest on a cold winter day?
    a. aluminum
    b. marble
    * c. wood
    d. iron

55. In conduction, thermal energy is transported by
    a. the movement of a fluid.
    * b. the collisions of particles.
    c. electromagnetic fields.
    d. the propagation of sound waves.

56. In convection, thermal energy is transported by
    * a. the movement of a fluid.
    b. the collisions of particles.
    c. electromagnetic fields.
    d. the propagation of sound waves.

57. In radiation, thermal energy is transported by
    a. the movement of a fluid.
    b. the collisions of particles.
    * c. electromagnetic fields.
    d. the propagation of sound waves.

Chapter 13   Thermal Energy

58. Which color star is the hottest?
       a. red
       b. yellow
   *   c. blue
       d. white

59. Many scientists are worried about the adverse effects of global warming brought on by the greenhouse effect. What causes the greenhouse effect?
       a. Green colored surfaces absorb more radiation than other colors.
       b. Building houses has adverse effects on the green revolution.
   *   c. Water vapor and carbon dioxide block infrared radiation.
       d. The increased use of greenhouses to grow crops out of season.

60. Which of the following does NOT affect the change in length of a bridge?
       a. length of the bridge
   *   b. cross-sectional area
       c. type of construction material
       d. change in the temperature

61. What is the change in length of a metal rod with an original length of 4 m, a coefficient of thermal expansion of $0.00002/^{o}C$, and a temperature change of $30^{o}C$?
       a. 24 mm
       b. 12 mm
   *   c. 2.4 mm
       d. 1.2 mm

62. A steel railroad rail is 24.4 m long. How much does it expand during a day when the low temperature is 50°F (18°C) and the high temperature is 91°F (33°C)? Steel has a coefficient of thermal expansion of $0.000011/^{o}C$
   *   a. 4 mm
       b. 4.8 mm
       c. 8.7 mm
       d. 9.6 mm

Chapter 14: Available Energy

1. It is NOT possible to completely convert
   - a. heat into internal energy.
   - b. mechanical energy into internal energy.
   - * c. internal energy into mechanical energy.
   - d. internal energy into heat.

2. A heat engine
   - * a. converts thermal energy into mechanical energy.
   - b. converts mechanical energy into thermal energy.
   - c. violates the first law of thermodynamics.
   - d. can be 100% efficient.

3. With his paddle-wheel apparatus James Joule determined that 4.2 joules of mechanical work were equivalent to 1 calorie of heat. If he had used a heat engine and had measured the heat flowing into the engine and the work done by the engine, he would have determined that
   - a. more than 4.2 joules were equivalent to 1 calorie.
   - * b. less than 4.2 joules were equivalent to 1 calorie.
   - c. 4.2 joules were still equivalent to 1 calorie.

4. The second law of thermodynamics says that
   - a. the energy of an isolated system is conserved.
   - * b. it is impossible to build a heat engine that can do mechanical work by extracting thermal energy that does not also exhaust heat to the surroundings.
   - c. it is impossible to reach the absolute zero of temperature.
   - d. it is impossible to build a heat engine that does more mechanical work than the thermal energy it consumes.

5. The second law of thermodynamics
   - a. is a restatement of the law of conservation of energy.
   - * b. says that heat cannot be completely converted to mechanical energy.
   - c. is the basis for the definition of temperature.
   - d. is the basis for the definition of internal energy.

6. It is impossible to run an ocean liner by taking in seawater at the bow of the ship, extracting internal energy from the water, and dropping ice cubes off the stern because this process violates the _____ law of thermodynamics.
   - a. zeroth
   - b. first
   - * c. second
   - d. third

7. Why is it impossible to run an ocean liner by taking in seawater at the bow of the ship, extracting internal energy from the water, and dropping ice cubes off the stern?
    a. This process violates the first law of thermodynamics.
* b. No heat engine can operate at a single temperature.
    c. Seawater does not contain enough energy.
    d. The temperature of seawater is too close to freezing.

8. An engine takes in 7000 cal of heat and exhausts 2000 cal of heat each minute it is running. How much work does the engine do each minute?
    a. 2000 cal
* b. 5000 cal
    c. 7000 cal
    d. 9000 cal

9. What input energy is required if an engine performs 50 kJ of work and exhausts 60 kJ of heat?
    a. 10 kJ
    b. 50 kJ
    c. 60 kJ
* d. 110 kJ

10. A heat engine requires an input of 5 kJ per minute to produce 2 kJ of work per minute. Therefore, it must exhaust _____ of heat per minute.
    a. 2 kJ
* b. 3 kJ
    c. 5 kJ
    d. 7 kJ

11. Many people have tried to build perpetual-motion machines. Our study of physics tells us that perpetual-motion machines are
    a. theoretically possible, but difficult to build.
    b. theoretically possible if they do not violate the first law of thermodynamics.
    c. theoretically possible, but would be very inefficient.
* d. not possible to build.

12. What restrictions does the first law of thermodynamics place on building a perpetual motion machine?
    a. It does not place any restriction on the possibility.
* b. Energy must be conserved.
    c. Thermal energy cannot be completely converted to mechanical energy.
    d. The machine must have a very long cyclical period.

13. What restrictions does the second law of thermodynamics place on building a perpetual motion machine?

      a. It does not place any restriction on the possibility.

      b. Energy must be conserved.

*   c. Thermal energy cannot be completely converted to mechanical energy.

      d. The machine must have a very long cyclical period.

14. Consider the human body to be a heat engine with an efficiency of 20%. This means that

      a. only 20% of the food you eat is digested.

      b. 80% of the energy you obtain from food is destroyed.

      c. you should spend 80% of each day lying quietly without working.

*   d. only 20% of the energy you obtain from food can be used to do work.

15. The efficiency of an ideal heat engine can be improved by _____ the input temperature and _____ the exhaust temperature.

      a. increasing ... increasing

*   b. increasing ... decreasing

      c. decreasing ... increasing

      d. decreasing ... decreasing

16. The efficiency of a real heat engine can be improved by _____ the input temperature and _____ the exhaust temperature.

      a. increasing ... increasing

*   b. increasing ... decreasing

      c. decreasing ... increasing

      d. decreasing ... decreasing

17. Nuclear power plants are less efficient than coal-fired plants because

      a. nuclear power plants have lower exhaust temperatures.

*   b. safety regulations require nuclear plants to run at lower temperatures.

      c. there is less energy in a ton of coal than a ton of uranium.

      d. coal plants produce much more carbon dioxide.

18. A heat engine takes in 600 J of energy at 1000 K and exhausts 300 J at 400 K. What is the actual efficiency of this engine?

      a. 25%

      b. 40%

*   c. 50%

      d. 75%

19. A heat engine takes in 600 J of energy at 1000 K and exhausts 300 J at 400 K. What is the maximum theoretical efficiency of this engine?
> a. 40%
> b. 50%
> * c. 60%
> d. 80%

20. A heat engine takes in energy at a rate of 1600 W at 1000 K and exhausts heat at a rate of 1200 W at 400 K. What is the actual efficiency of this engine?
> * a. 25%
> b. 40%
> c. 50%
> d. 75%

21. A heat engine takes in energy at a rate of 1600 W at 1000 K and exhausts heat at a rate of 1200 W at 400 K. What is the maximum theoretical efficiency of this engine?
> a. 40%
> b. 50%
> * c. 60%
> d. 70%

22. When the input to an engine is 1000 W at 800 K and the exhaust temperature is 400 K, the engine performs work at a rate of 200 W. At what rate does the engine exhaust heat?
> a. 200 W
> b. 400 W
> c. 600 W
> * d. 800 W

23. When the input to an engine is 1000 W at 800 K and the exhaust temperature is 400 K, what is the minimum theoretical rate the engine could exhaust heat?
> a. 200 W
> b. 400 W
> * c. 500 W
> d. 600 W

24. A heat engine operating at an efficiency of 20% performs 500 J of work per second. What is the input power to the engine?
> a. 100 W
> b. 600 W
> c. 2000 W
> * d. 2500 W

25. A heat engine operating at an efficiency of 20% performs 500 J of work per second. What is the exhaust rate?
    a. 400 W
    b. 900 W
    * c. 2000 W
    d. 2500 W

26. An engine has an efficiency of 40%. How much energy must be extracted to do 900 J of work?
    a. 360 J
    b. 900 J
    * c. 2250 J
    d. 3600 J

27. An engine exhausts 1200 J of energy for every 3600 J of energy it takes in. What is its efficiency?
    a. 25 %
    b. 33 %
    c. 50 %
    * d. 67 %

28. An engineer has designed a machine to produce electricity by using the difference in the temperature of ocean water at different depths. If the surface temperature is 20°C and the temperature at 50 m below the surface is 12°C, what is the maximum efficiency of this machine?
    * a. 3 %
    b. 40 %
    c. 60 %
    d. 97 %

29. An ideal heat engine has a theoretical efficiency of 60% and an exhaust temperature of 27°C. What is its input temperature?
    a. 227 °C
    * b. 477 °C
    c. 500 °C
    d. 750 °C

30. What is the exhaust temperature of an ideal heat engine that has an efficiency of 50% and an input temperature of 500°C?
    * a. 114 °C
      b. 250 °C
      c. 387 °C
      d. 1000 °C

31. The second law of thermodynamics says that
    a. the energy of an isolated system is conserved.
    * b. it is impossible to build a refrigerator that can transfer heat from a lower temperature region to a higher temperature region without expending mechanical work.
      c. it is impossible to reach the absolute zero of temperature.
      d. it is impossible to build a refrigerator that exhausts more heat than the mechanical work required to operate it.

32. An air-conditioner mechanic is testing a unit by running it on the workbench in an isolated room? What happens to the temperature of the room?
    * a. It increases.
      b. It decreases.
      c. It stays the same.

33. The coefficient of performance for a refrigerator is defined as the ratio of the heat extracted from the colder system to the work required. The coefficient of performance for a good refrigerator is
    a. less than one.
    b. equal to one.
    * c. greater than one.

34. The coefficient of performance of an ideal refrigerator can be improved by _____ the input temperature and _____ the exhaust temperature.
    a. increasing ... increasing
    * b. increasing ... decreasing
      c. decreasing ... increasing
      d. decreasing ... decreasing

35. The coefficient of performance of a real refrigerator can be improved by _____ the input temperature and _____ the exhaust temperature.
    a. increasing ... increasing
    * b. increasing ... decreasing
      c. decreasing ... increasing
      d. decreasing ... decreasing

36. A refrigerator extracts 1000 J of energy from a cold region and exhausts 1400 J of energy to a hot region. How much work was required?
    * a. 400 J
    b. 1000 J
    c. 1400 J
    d. 2400 J

37. A refrigerator requires 400 J to extract 1000 J of energy from a cold region. How much energy does it exhaust to a hot region?
    a. 400 J
    b. 600 J
    c. 1000 J
    * d. 1400 J

38. How much work per second (power) is required by a refrigerator that takes 800 J of thermal energy from a cold region each second and exhausts 1200 J each second to a hot region?
    * a. 400 W
    b. 800 W
    c. 1200 W
    d. 2000 W

39. A refrigerator requires mechanical energy at a rate of 200 W to extract thermal energy from a cold region at a rate of 1000 W. What is the coefficient of performance for this refrigerator?
    a. 0.2
    b. 2
    * c. 5
    d. 6

40. What is the maximum coefficient of performance for a refrigerator operating between temperatures of 200 K and 300 K?
    a. 1/2
    b. 2/3
    c. 3/2
    * d. 2

41. If a refrigerator requires an input of 200 J of electrical energy each second and has a coefficient of performance of 5, how much heat energy is extracted from the refrigerator each second?
    a. 40 J
    b. 100 J
    c. 800 J
    * d. 1000 J

42. How many different arrangements can you make with three different colored blocks?
         a. 3
    *   b. 6
         c. 9
         d. 12

43. How many different arrangements are there for four different coins?
         a. 4
         b. 8
         c. 12
    *   d. 16

44. What is the probability of flipping three coins and obtaining 2 heads and a tail?
         a. 1/8
         b. 2/8
    *   c. 3/8
         d. 4/8

45. What is the probability of throwing two dice and obtaining a sum of seven?
         a. 1/36
         b. 2/36
         c. 3/36
    *   d. 6/36

46. What is the probability of rolling a total of 6 with two dice?
         a. 1/36
         b. 3/36
    *   c. 5/36
         d. 6/36

47. What is the probability of rolling a total of 10 with two dice?
         a. 1/36
    *   b. 3/36
         c. 5/36
         d. 10/36

48. Given a system of three identical coins, how many equivalent states have two tails and one head?
         a. 1
         b. 2
    *   c. 3
         d. 4

49. Which law of thermodynamics is the basis for the definition of entropy?
    a. zeroth
    b. first
    * c. second
    d. third

50. The second law of thermodynamics
    a. says that it is impossible to reach the absolute zero of temperature.
    * b. says that the entropy of a system tends to increase.
    c. is the basis for the definition of temperature.
    d. is the basis for the definition of internal energy.

51. Which of the following disagrees with the second law of thermodynamics?
    a. Heat naturally flows from hot objects to cold objects.
    b. No engine can transform all of its heat input into mechanical work.
    * c. The entropy of an isolated system can never decrease.
    d. Perpetual motion machines are not possible.

52. A ringing bell is inserted into a large glass of water. The bell and the water are initially at the same temperature and are insulated from their surroundings. Eventually the bell stops vibrating and the water comes to rest. Which of the following statements is FALSE?
    a. The mechanical energy of the bell has been completely converted into internal energy of the combined system.
    b. The final temperature of the combined system is lower than the initial temperature.
    c. The entropy of the combined system has increased.
    * d. None of the statements is false.

53. A bell is placed in a glass of hot water and then removed. Which of the following statements is FALSE?
    * a. The bell will be ringing.
    b. The bell will be warmer.
    c. Energy has been transferred from the hot water to the bell.
    d. The entropy of the system has increased.

54. A ball lying on the floor does NOT coordinate the kinetic energy of its randomly moving atoms and suddenly leap up off the floor because
    a. this would violate the first law of thermodynamics.
    b. this would violate the second law of thermodynamics.
    c. the total energy of the system would have to increase.
    * d. Although it could happen in principle, it is extremely unlikely.

55. As a baby grows into an adult, the entropy of the person
      a. increases.
      b. stays the same.
  * c. decreases.

56. In which of the systems listed below is the entropy decreasing?
  * a. A gas is cooled.
      b. A plate is shattered.
      c. An egg is scrambled.
      d. A drop of dye diffuses in a cup of water.

57. A cold piece of metal is dropped into an insulated container of hot water. After the system has reach an equilibrium temperature, the
      a. entropy of the metal has decreased.
  * b. entropy of the water has decreased.
      c. net change in entropy of the system is zero.
      d. entropy of the system has decreased.

58. A cold piece of metal is dropped into an insulated container of hot water. After the system has reach an equilibrium temperature, the
      a. entropy of the metal has decreased.
      b. entropy of the water has increased.
      c. net change in entropy of the system is zero.
  * d. entropy of the system has increased.

59. As a baby grows into an adult, the entropy of the Universe
  * a. increases.
      b. stays the same.
      c. decreases.

60. We are currently experiencing a world-wide energy crisis because the
      a. amount of energy in the world is decreasing rapidly.
  * b. entropy of the world is increasing rapidly.
      c. entropy of the world is decreasing rapidly.
      d. amount of energy in the world is not increasing.

Chapter 15: Vibrations and Waves

1. Which of the following statements is NOT always true for a mass suspended from a vertical spring? When the mass is above the equilibrium position, the _____ points downward.
   - a. restoring force
   - b. force of gravity
   - c. acceleration
   - * d. velocity

2. If a cycle begins at a certain position, it ends when the object
   - a. next returns to this position.
   - b. next travels in the same direction.
   - * c. next returns to this position while traveling in the same direction.
   - d. passes through the equilibrium position for the second time.

3. For small amplitudes the restoring force on a mass on a spring is _____ its distance from the equilibrium position.
   - * a. proportional to
   - b. proportional to the square of
   - c. proportional to the square root of
   - d. inversely proportional to

4. What is the period of the hand on a clock that measures the seconds?
   - a. 1/60 s
   - b. 1 s
   - * c. 60 s
   - d. 3600 s

5. What is the period of the hand on a clock that measures the minutes?
   - a. 1 second
   - b. 1 minute
   - * c. 1 hour
   - d. 12 hours

6. A mass suspended from a spring is seen to bob up and down over a distance of 20 cm from the top to the bottom of its path twice each second. What is its amplitude?
   - * a. 10 cm
   - b. 20 cm
   - c. 2 s
   - d. 0.5 Hz

Chapter 15   Vibrations and Waves

7. A bob vibrates up and down on a vertical spring. If the period of the motion is two seconds, the frequency is
        a. also two seconds.
        b. two cycles per second.
        c. one cycle per second.
   *   d. one-half cycle per second.

8. A bob oscillates on a vertical spring. If the frequency of the motion is two cycles per second, the period is
        a. also two cycles per second.
        b. one second.
        c. two seconds.
   *   d. one-half second.

9. If a mass on the end of a spring takes 3 s to complete one cycle, its frequency is
        a. 3 s.
        b. 1/3 s.
        c. 3 Hz.
   *   d. 1/3 Hz.

10. A short pendulum has a frequency of 2 Hz, its period is
        a. 2 s.
   *   b. 1/2 s.
        c. 2 cycles/s.
        d. 1/2 cycle/s.

11. A long pendulum hanging from the ceiling of the classroom has a period of 4 s. What is the frequency of this pendulum?
        a. 4 Hz
        b. 2 Hz
        c. 1/2 Hz
   *   d. 1/4 Hz

12. What is the frequency of the hand on a clock that measures the seconds?
        a. 60 Hz
        b. 1 Hz
   *   c. 1/60 Hz
        d. 1/3600 Hz

13. For small amplitudes the period for the motion of an object on a spring is _____ the mass of the object.
   a. proportional to
 * b. proportional to the square root of
   c. inversely proportional to the square root of
   d. inversely proportional to

14. For small amplitudes the period for the motion of an object on a spring is _____ the value of the spring constant.
   a. proportional to
   b. proportional to the square root of
 * c. inversely proportional to the square root of
   d. inversely proportional to

15. Increasing which of the following increases the period of a mass on a spring?
 * a. mass
   b. amplitude
   c. spring constant
   d. strength of gravity

16. Increasing which of the following decreases the period of a mass on a spring?
   a. mass
   b. amplitude
 * c. spring constant
   d. strength of gravity

17. A spring oscillates with a period of 1 s with a mass of 0.25 kg. What would its period be if the mass were increased to 1 kg?
   a. 0.5 s
   b. 1 s
 * c. 2 s
   d. 4 s

18. Changing which of the following affects the period of a pendulum?
   a. mass
   b. amplitude
 * c. length
   d. angle

Chapter 15   Vibrations and Waves

19. Increasing which of the following causes an increase in the period of a pendulum?
    a. mass
    b. amplitude
  * c. length
    d. acceleration due to gravity

20. Increasing which of the following causes a decrease in the period of a pendulum?
    a. mass
    b. amplitude
    c. length
  * d. acceleration due to gravity

21. A pendulum is allowed to swing through one cycle and its period is measured. It is now restarted with twice the amplitude. The new period is
    a. twice as big.
    b. half as big.
  * c. the same.

22. For small amplitudes the period of a pendulum is _____ the length of the string.
    a. proportional to
  * b. proportional to the square root of
    c. inversely proportional to the square root of
    d. inversely proportional to

23. For small amplitudes the period of a pendulum is _____ the acceleration due to gravity.
    a. proportional to
    b. proportional to the square root of
  * c. inversely proportional to the square root of
    d. inversely proportional to

24. A pendulum with a length of 2 m has a period of 2.8 s. What is the period of a pendulum with a length of 8 m?
    a. 1.4 s
    b. 2.8 s
  * c. 5.6 s
    d. 11.2 s

25. A grandfather clock runs too slow. Which of the following could you do to correct this?
    a. remove some mass from the bob
    b. increase the amplitude
    c. move the bob down.
  * d. move the bob up.

26. What caused the Tacoma Narrows Bridge to collapse?
    a. The wind was strong enough to blow the bridge down.
    b. The frequency of the wind matched a natural frequency of the bridge.
  * c. The frequency of the turbulence due to the steady wind matched the resonant frequency of the bridge.
    d. The bridge was not strong enough to support the high volume of traffic.

27. Why do soldiers break step before crossing a suspension bridge?
    a. They cannot walk as fast on the bridge.
  * b. Their cadence might match the natural frequency of the bridge.
    c. Because of long standing tradition passed down from Hannibal.
    d. Superstition

28. Which of the following multiples of the fundamental frequency will not produce resonance for a child's swing?
    a. 1/3
    b. 1/2
    c. 1
  * d. 2

29. A short pendulum has a natural frequency of 4 Hz. Which of the following driving frequencies would not produce resonance for this pendulum?
    a. 1 Hz
    b. 2 Hz
    c. 4 Hz
  * d. 8 Hz

30. When a wave travels from one place to another, what is transported?
    a. mass of the medium
    b. mass from the source
  * c. energy
    d. nothing

31. The waves that occur on a clothesline are _____ waves.
    a. particle
  * b. transverse
    c. longitudinal
    d. tangential

Chapter 15   Vibrations and Waves

32. The waves that occur in fluids are _____ waves.
    a. particle
    b. transverse
    * c. longitudinal
    d. tangential

33. A wave in which the individual particles of a medium vibrate back and forth perpendicular to the direction the wave travels is called a _____ wave.
    a. particle
    * b. transverse
    c. harmonic
    d. longitudinal

34. A wave in which the individual particles of a medium vibrate back and forth parallel to the direction the wave travels is called a _____ wave.
    a. particle
    b. transverse
    c. harmonic
    * d. longitudinal

35. Two wave pulses are sent down a very long rope. The first wave is smaller than the second one. Which of the following is true? The bigger wave will
    a. not catch the smaller one because the bigger wave has a smaller speed.
    b. catch the smaller one because the bigger wave has a larger speed.
    * c. not catch the smaller one because wave speeds are independent of size.

36. You observe the wave shape shown above traveling along a rope. Which of the following shapes would you expect to see after the wave reflects from an end that's securely fixed to a post?

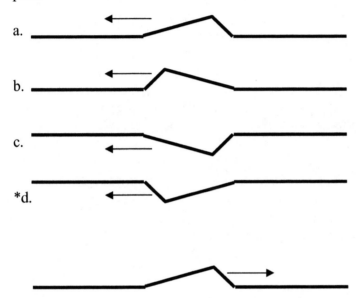

a.

b.

c.

*d.

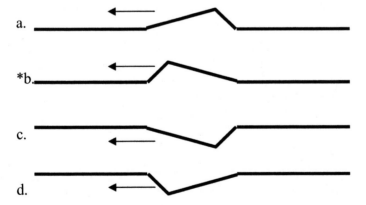

37. You observe the wave shape shown above traveling along a rope. Which of the following shapes would you expect to see after the wave reflects from an end that's free?

a.

*b.

c.

d.

Chapter 15   Vibrations and Waves

38. Which of the following affects the speed of a wave pulse on a rope?
       a. length of the rope
       b. length of the pulse
       c. amplitude of the pulse
   *   d. tension in the rope

39. Two wave pulses are sent in opposite directions down a long rope. If they are both "crests," we observe that when they meet, they
   *   a. pass through each other.
       b. cancel each other.
       c. bounce off each other and become troughs.
       d. bounce off each other and remain crests.

40. The distance from trough to trough on a periodic wave is called its
       a. frequency.
       b. period.
   *   c. wavelength.
       d. amplitude.

41. The velocity of a periodic wave is equal to its wavelength
   *   a. times its frequency.
       b. divided by its frequency.
       c. times its period.
       d. divided by its amplitude.

42. Which of the following expressions gives the correct relationship between the wavelength, the period or frequency, and the velocity for a periodic wave?
       a. $v = \lambda T$
   *   b. $v = \lambda f$
       c. $v = \lambda/f$
       d. $v = fT$

43. How far does a wave travel during one period?
   *   a. a wavelength
       b. a crest
       c. a trough

44. In the following list of properties of periodic waves, which one is independent of the others?
       a. frequency
       b. wavelength
       c. speed
   *   d. amplitude

45. Two waves with the same speed have different wavelengths. Which has the lower frequency? The one with the
    * a. longer wavelength.
    b. shorter wavelength.
    c. smaller amplitude.
    d. larger amplitude.

46. As you increase the frequency of putting your finger into water, the wavelength of the water waves
    a. increases.
    * b. decreases.
    c. stays the same.

47. Assume that you have a periodic wave traveling down an infinitely long rope. What happens to the wavelength if you increase the period of this wave?
    a. Nothing, it stays the same.
    * b. It increases.
    c. It decreases.

48. If we compare the speed of a periodic sound wave with a frequency of 220 Hz to that of a wave with a frequency of 440 Hz, the 220-Hz wave is moving _____ as the 440-Hz wave.
    a. twice as fast
    b. half as fast
    * c. at the same speed

49. If the frequency of a periodic wave is doubled while the velocity remains the same, the wavelength
    a. doubles.
    b. quadruples.
    c. stays the same.
    * d. is cut in half.

50. A periodic wave on a string has a wavelength of 30 cm and a frequency of 2 Hz. What is the speed of the wave?
    a. 15 cm/s
    b. 30 cm/s
    * c. 60 cm/s
    d. 120 cm/s

51. If the speed of a 25-Hz wave doubles, the wavelength
    * a. doubles.
      b. quadruples.
      c. stays the same.
      d. is cut in half.

52. While standing on a dock, you observe water waves passing beneath you. If you observe 20 wave crests passing directly below you in 2 minutes and judge the crests to be 1 m apart, what is the period of the waves?
      a. 20 s
      b. 10 s
    * c. 6 s
      d. 2 s

53. While standing on a dock, you observe water waves passing beneath you. If you observe 20 wave crests passing directly below you in 2 minutes and judge the crests to be 1 m apart, what is the speed of the waves?
    * a. 0.17 m/s
      b. 10 m/s
      c. 20 m/s
      d. 40 m/s

54. If the frequency of a wave is 2 Hz and there are 3 m between adjacent crests, how far does a crest travel in 2 s?
      a. 1.5 m
      b. 3 m
      c. 6 m
    * d. 12 m

55. Standing waves
    * a. are a result of the superposition of two traveling waves.
      b. can only occur in vibrating strings.
      c. only occur when the system is at certain temperatures.
      d. move with a speed equal to the speed of sound.

56. A node of a standing wave is a
      a. spot of maximum amplitude.
    * b. spot of minimum amplitude.
      c. traveling wave.
      d. note that's an octave higher than the fundamental.

57. Two standing waves are created on a rope. One has a shorter wavelength than the other. The frequency of the one with the shorter wavelength will be _____ the frequency of the one with the longer wavelength.
    - a. equal to
    - b. smaller than
    - * c. larger than

58. Increasing the tension in the rope _____ the fundamental frequency of a rope.
    - * a. increases
    - b. decreases
    - c. does not change

59. The fundamental wavelength for standing waves on a rope fixed at both ends is ____ the length of the rope.
    - a. four times
    - * b. two times
    - c. the same as
    - d. one-half

60. The wavelength of the second harmonic for standing waves on a rope fixed at both ends is ____ the length of the rope.
    - a. four times
    - b. two times
    - * c. the same as
    - d. one-half

61. If the fundamental frequency for a 2-m-long rope fixed at both ends is 10 Hz, the speed of the wave is
    - * a. 40 m/s.
    - b. 20 m/s.
    - c. 10 m/s.
    - d. 5 m/s.

62. If the fundamental frequency of a rope fixed at both ends is 30 Hz, the frequency of the second harmonic is
    - a. 15 Hz.
    - b. 20 Hz.
    - c. 30 Hz.
    - * d. 60 Hz.

63. The frequency of the third harmonic on a rope fixed at both ends is _____ times that of the fundamental frequency.
    a. 5
  * b. 3
    c. 2
    d. 1/3

64. Two point sources produce waves of the same wavelength and are completely in-phase (that is, both sources produce crests at the same time). At a point midway between the sources, you would expect to find an oscillation with
  * a. an amplitude equal to twice that of one wave alone.
    b. an amplitude equal to that of one wave alone.
    c. approximately zero amplitude.

65. Two point sources produce waves of the same wavelength and are completely in-phase (that is, both sources produce crests at the same time). At a point that is one wavelength farther from one source than the other, you would expect to find an oscillation with
  * a. an amplitude equal to twice that of one wave alone.
    b. an amplitude equal to that of one wave alone.
    c. approximately zero amplitude.

66. A two-slit interference pattern is produced in a ripple tank. As the two probes are brought closer together, the separation of the locations of maximum amplitude along the far edge of the tank
    a. decreases.
  * b. increases.
    c. remains the same.

67. A two-slit interference pattern is produced in a ripple tank. As the frequency of the probes increases, the separation of the locations of maximum amplitude along the far edge of the tank
  * a. decreases.
    b. increases.
    c. remains the same.

68. A single-slit diffraction pattern is produced in a ripple tank. As the slit is made wider, the separation of the locations of maximum amplitude along the far edge of the tank
  * a. decreases.
    b. increases.
    c. remains the same.

Physics: A World View, Sixth Edition by Larry Kirkpatrick and Gregory Francis

69. A single-slit diffraction pattern is produced in a ripple tank. As the wavelength is made shorter, the separation of the locations of maximum amplitude along the far edge of the tank
*   a. decreases.
     b. increases.
     c. remains the same.

Chapter 16: Sound and Music

1. When a sound wave travels from one place to another, what is transported?
    - a. air
    - b. density
    - * c. energy
    - d. nothing

2. Sound waves are _____ waves.
    - a. transverse
    - * b. longitudinal
    - c. oblique
    - d. particle

3. Sonar waves are _____ waves.
    - a. transverse
    - * b. longitudinal
    - c. oblique
    - d. particle

4. Which of the following is NOT true? The speed of sound in air
    - a. is slower than in water.
    - b. depends on the temperature.
    - c. depends on the type of gas.
    - * d. depends upon the source.

5. Which of the following has the most effect on the speed of sound in air?
    - a. amplitude
    - b. frequency
    - c. wavelength
    - * d. temperature

6. In which of the following is the speed of sound the fastest?
    - a. water
    - b. helium
    - c. air
    - * d. iron

7. You see the lightning before you hear the thunder because
    - a. the speed of sound is faster than that of light.
    - * b. the speed of light is faster than that of sound.
    - c. the thunder forms after the lightning.
    - d. your eyes are more sensitive to light.

8. Even though you may be far away from an orchestra, the tuba and the piccolo do not sound "out of step" with each other. This indicates that sound waves
    a. are longitudinal.
    b. are transverse.
    c. tend to harmonize.
*   d. travel at the same speed for all frequencies.

9. If we compare the speed of a periodic sound wave with a frequency of 220 Hz to that of a wave with a frequency of 440 Hz, the 220-Hz wave is moving _____ as the 440-Hz wave.
    a. twice as fast
    b. half as fast
*   c. at the same speed

10. If you have to wait 10 seconds between the arrival of the flash and the thunder, you know that the lightning bolt occurred _____ away.
    a. 1/10 mile
*   b. 2 miles
    c. 5 miles
    d. 10 miles

11. If you have to wait 9 seconds between the arrival of the flash and the thunder, you know that the lightning bolt occurred _____ away.
    a. 1/9 kilometer
    b. 1 kilometer
*   c. 3 kilometers
    d. 9 kilometers

12. The amplitude of a sound wave is closely related to its
*   a. loudness.
    b. pitch.
    c. wavelength.
    d. harmonics.

13. The loudness (or intensity) of a sound wave is related to its
    a. duration.
    b. frequency.
    c. wavelength.
*   d. amplitude.

14. The pitch of a sound wave is primarily determined by its
  a. amplitude.
  * b. frequency.
  c. speed.
  d. harmonics.

15. A frequency that is _____ times the fundamental frequency will be two octaves higher.
  a. one-half
  b. two
  c. three
  * d. four

16. Which of the following harmonics is two octaves higher in pitch than the fundamental?
  a. first
  b. second
  c. third
  * d. fourth

17. Which of the following harmonics is one octave lower than the fundamental?
  a. second
  b. third
  c. fourth
  * d. None of these

18. Two standing waves are created on a guitar string. One has a longer wavelength than the other. The frequency of the one with the longer wavelength will be _____ the frequency of the one with the longer wavelength.
  a. equal to
  * b. smaller than
  c. larger than

19. The length of a guitar's fundamental wavelength is
  * a. longer than the string.
  b. shorter than the string.
  c. equal to the length of the string.

20. Where on a guitar string of length L would you place your finger to damp out the fourth harmonic?
  a. middle
  b. 1/4 L from either end
  * c. 1/8 L from either end
  d. 1/16 L from either end

21. How does increasing the tension of the string affect the fundamental frequency of a guitar string? The frequency
    * a. increases.
    b. decreases.
    c. stays the same.

22. How does plucking the string harder affect the fundamental frequency of a guitar string? The frequency
    a. increases.
    b. decreases.
    * c. stays the same.

23. Which of the following actions will decrease the frequency of the note played on a guitar string?
    a. decreasing the mass of the string
    * b. decreasing the tension in the string
    c. fingering the string
    d. plucking the string harder

24. Which of the following actions will increase the frequency of the note played on a guitar string?
    a. increasing the mass of the string
    b. decreasing the tension in the string
    * c. fingering the string
    d. plucking the string harder

25. Which of the following would you expect to find at the end of a guitar string?
    a. nodule
    * b. node
    c. antimony
    d. antinode

26. The wavelength corresponding to the fundamental frequency of a guitar string is _____ times the length of the string.
    a. 1/4
    b. 1/2
    c. 1
    * d. 2

Chapter 16   Sound and Music

27. The wavelength corresponding to the third harmonic of a guitar string is _____ times the length of the string.
    a. 1/3
* b. 2/3
    c. 1
    d. 3

28. If the fundamental frequency of a guitar string is 220 Hz, the frequency of the second harmonic is
    a. 110 Hz.
    b. 220 Hz.
* c. 440 Hz.
    d. 880 Hz.

29. The frequency of the third harmonic on a guitar string is _____ times that of the fundamental frequency.
    a. 5
* b. 3
    c. 2
    d. 1/3

30. The fundamental wavelength and frequency for a guitar string are 1.2 m and 240 Hz, respectively. What is the speed of traveling waves on the string?
    a. 120 m/s
    b. 200 m/s
* c. 288 m/s
    d. 360 m/s

31. If the wave speed on a 50-cm long wire fixed at both ends is 300 m/s, what is the frequency of the fundamental?
    a. 150 Hz
* b. 300 Hz
    c. 600 Hz
    d. 1200 Hz

32. If the wave speed on a 25-cm long wire fixed at both ends is 400 m/s, what is the frequency of the fundamental?
    a. 100 Hz
    b. 200 Hz
    c. 400 Hz
* d. 800 Hz

Physics: A World View, Sixth Edition by Larry Kirkpatrick and Gregory Francis

33. If the fundamental frequency of a 60-cm-long guitar string is 500 Hz, what is the speed of the traveling waves on the string?
     a. 300 m/s
*   b. 600 m/s
     c. 833 m/s
     d. 1200 m/s

34. A longitudinal standing wave can be established in a long, aluminum rod by stroking it with rosin on your fingers. If the rod is held midway between the two ends of the rod, the wavelength of the fundamental will be ____ the length of the rod.
     a. one quarter
     b. one half
     c. equal to
*   d. twice

35. A longitudinal standing wave can be established in a long, aluminum rod by stroking it with rosin on your fingers. If the rod is held midway between the center and one end of the rod, the wavelength of the fundamental will be ____ the length of the rod.
     a. one quarter
     b. one half
*   c. equal to
     d. twice

36. A closed organ pipe has
*   a. a node at the closed end and an antinode at the open end.
     b. an antinode at the closed end and a node at the open end.
     c. a node at each end.
     d. an antinode at each end.

37. If you saw a bit off the end of an organ pipe, the fundamental wavelength will
     a. increase.
*   b. decrease.
     c. stay the same.

38. If you saw a bit off the end of an organ pipe, the fundamental frequency will
*   a. increase.
     b. decrease.
     c. stay the same.

39. If you fill an organ pipe with helium instead of air, the fundamental wavelength will
    a. increase.
    b. decrease.
 * c. stay the same.

40. If you fill an organ pipe with helium instead of air, the fundamental frequency will
 * a. increase.
    b. decrease.
    c. stay the same.

41. Which of the following would you expect to find near the open end of a trumpet?
    a. node
 * b. antinode
    c. crest
    d. trough

42. Which of the following musical instruments does not have even harmonics?
    a. guitar
    b. open organ pipe
 * c. closed organ pipe
    d. harp

43. The wavelength of the fundamental standing wave for an organ pipe that is open at both ends is _____ the length of the pipe.
    a. one-half
    b. the same as
 * c. two times
    d. four times

44. The wavelength of the fundamental standing wave for an organ pipe that is open at one end and closed at the other is _____ the length of the pipe.
    a. one-half
    b. the same as
    c. two times
 * d. four times

45. If you close one end of an open organ pipe, the wavelength of the fundamental
    a. is cut in half.
    b. stays the same.
 * c. doubles.

46. If you close one end of an open organ pipe, the frequency of the fundamental
    *   a. is cut in half.
        b. stays the same.
        c. doubles.

47. If you open the closed end of a closed organ pipe, the frequency of the fundamental
        a. is cut in half.
        b. stays the same.
    *   c. doubles.

48. If you open the closed end of a closed organ pipe, the wavelength of the fundamental
    *   a. is cut in half.
        b. stays the same.
        c. doubles.

49. What length of closed organ pipe is required to produce the note B$_4$ with a frequency of 494 Hz? Assume that it is a sunny spring day and the speed of sound is 343 m/s.
    *   a. 0.17 m
        b. 0.69 m
        c. 1.4 m
        d. 2.8 m

50. You have an organ pipe that resonates at frequencies of 500, 700, and 900 Hz but nothing in-between. It may resonate at lower and higher frequencies as well. What is the fundamental frequency for this pipe?
    *   a. 100 Hz
        b. 200 Hz
        c. 500 Hz

51. You have an organ pipe that resonates at frequencies of 500, 700, and 900 Hz but nothing in-between. It may resonate at lower and higher frequencies as well. This pipe is
        a. open at both ends.
    *   b. open at one end and closed at the other.
        c. closed at both ends.

52. You have an organ pipe that resonates at frequencies of 375, 450, and 525 Hz but nothing in-between. It may resonate at lower and higher frequencies as well. What is the fundamental frequency for this pipe?
    *   a. 75 Hz
        b. 150 Hz
        c. 375 Hz

53. The fundamental wavelength for a bar on a xylophone that is suspended at points midway between the ends and the middle of the bar is ____ times the length of the bar.
    - a. 1/4
    - b. 1/2
    - * c. 1
    - d. 2

54. Even though a guitar and a banjo play the same note, they sound different because the
    - a. musician can play one instrument better than the other.
    - b. guitar is more expensive.
    - c. notes do not have the same frequency.
    - * d. intensities of the harmonics differ.

55. The two wires corresponding to one key on a piano are out of tune. If we increase the tension of the wire producing the higher frequency, the two frequencies produce a beat frequency that
    - a. becomes lower.
    - * b. becomes higher.
    - c. stays the same.

56. The two wires corresponding to one key on a piano are out of tune. If we decrease the tension of the wire producing the lower frequency, the two frequencies produce a beat frequency that
    - a. becomes lower.
    - * b. becomes higher.
    - c. stays the same.

57. As the frequencies of two waves get closer together, the beat frequency
    - a. increases.
    - b. stays the same.
    - * c. decreases.

58. Beats occur because
    - a. waves reflect off fixed boundaries.
    - b. harmonics interfere destructively.
    - * c. waves exhibit superposition.
    - d. frequencies shift when the source of the sound moves.

59. What is the beat frequency of two waves with frequencies of 348 Hz and 352 Hz?
    - a. 2 Hz
    - * b. 4 Hz
    - c. 8 Hz
    - d. 350 Hz

60. Two piano wires produce a beat frequency of 2 Hz. If one wire is known to have a frequency of 440 Hz, what is the frequency of the other wire?
    a. 438 Hz
    * b. 438 Hz or 442 Hz
    c. 442 Hz
    d. 444 Hz

61. You have a tuning fork of unknown frequency. When you ring it alongside a tuning fork with a known frequency of 360 Hz, you hear beats at a frequency of 2 Hz. When you ring it alongside a tuning fork with a known frequency of 355 Hz, you hear beats at a frequency of 3 Hz. What is the unknown frequency?
    a. 352 Hz
    b. 357 Hz
    * c. 358 H
    d. 362 H

62. When a siren of frequency 400 Hz moves away from you with a constant speed, the frequency of the sound you hear
    a. continually increases as it gets farther away.
    b. continually decreases as it gets farther away.
    * c. is lower than 400 Hz and constant.
    d. is higher than 400 Hz and constant.

63. Which of the following properties of the wave does not change in the Doppler effect?
    a. wavelength
    * b. speed
    c. frequency
    d. They all change.

64. The frequency emitted by a police radar gun is 10 GHz. When you drive your car toward a state patrol car with a constant speed, the frequency of the radar returning to the patrol car
    a. continually increases as you get closer.
    b. continually decreases as you get closer.
    * c. is higher than 10 GHz and constant.
    d. is lower than 10 GHz and constant.

65. A police car's siren has a frequency of 1000 Hz. In which of the following will an observer hear a higher frequency?
    * a. The police car is moving <u>toward</u> a stationary observer.
    b. The police car is moving <u>away from</u> a stationary observer.
    c. Both the police car and the observer are <u>stationary</u>.
    d. The observer is running <u>away from</u> a stationary police car.

66. An automobile sounding its horn is moving away from you. The pitch (frequency) of the horn that you hear is _____ the pitch heard by the driver.
      a. higher than
*   b. lower than
      c. the same as

67. Billy and Elaine are holding electric buzzers that sound at slightly different frequencies. When they are both standing still, the two buzzers produce a beat frequency of 4 hertz. Elaine begins to run away, and Billy hears the beat frequency gradually increase. Whose buzzer has the higher frequency?
      a. Billy's
*   b. Elaine's
      c. The frequencies are the same.

68. The sonic boom occurs
      a. only when the airplane passes through the speed of sound.
      b. when the airplane is moving at slightly less than the speed of sound.
*   c. during the entire time the airplane exceeds the speed of sound.
      d. There is no such thing as a sonic boom.

69. The observation of a wake created by a boat tells you that the speed of the boat is _____ that of water waves.
*   a. greater than
      b. smaller than
      c. the same as

70. You go out and spend a bucket of money on a new stereo amplifier that advertises that it will be able to produce maximum volumes 20 dB greater than your current system. How many times more powerful is this amplifier?
      a. 20
*   b. 100
      c. 200
      d. 400

71. If ear protectors can reduce the sound intensity by a factor of 10,000, by how many decibels is the sound level reduced?
      a. 0.0001 dB
      b. 4 dB
*   c. 40 dB
      d. 10,000 dB

Chapter 17: Light

1. A laser beam is shone across the classroom. While the spot on the wall is visible to you, you can't see the beam between the laser and the wall. This is because light
    a. needs a vacuum to be visible.
  * b. needs to enter your eyes to be visible.
    c. needs to chemically react with the wall to be visible.
    d. does not turn red until it hits the wall.

2. Why can you usually see the beam of a searchlight pass through the air?
    a. The vacuum makes its visible.
    b. The light reacts chemically with the pollutants in the air.
    c. The light emitted by your eyes interacts with the beam.
  * d. Some of the light is scattered into your eyes by particles in the air.

3. When you look at a crescent moon on a clear night, you can often see the rest of the face of the Moon very dimly lit. This occurs because
    a. The Moon glows from residual radioactivity in its rocks.
    b. The Moon produces its own light.
    c. Some of the sunlight passes through the Moon.
  * d. Some light is reflected form Earth and strikes the Moon.

4. Which of the following will cast a shadow that has an umbra but no penumbra?
    a. the Sun
    b. an incandescent lamp
    c. a fluorescent lamp
  * d. a point source of light

5. The shadow of a hand produced by the beam from a slide projector (a point source of light) has
    a. a penumbra but no umbra.
  * b. an umbra but no penumbra.
    c. an umbra and a penumbra.
    d. neither an umbra nor a penumbra.

6. A total eclipse of the Sun occurs when
    a. you stand in the penumbra of the Moon's shadow.
  * b. you stand in the umbra of the Moon's shadow.
    c. sunlight diffracts around the Moon.
    d. sunlight reflects from the Moon to Earth.

7. During a partial solar eclipse, the shadow produced on Earth's surface has
    *   a. a penumbra but no umbra.
         b. an umbra but no penumbra.
         c. an umbra and a penumbra.
         d. neither an umbra nor a penumbra.

8. You hold your hand 3 feet above the ground and look at the shadow cast by the Sun. You repeat this inside using the light from a 2-foot by 4-foot fluorescent light box in the ceiling. In which case will the penumbra be more pronounced?
         a. the Sun
    *   b. the fluorescent light box
         c. Both will produce the roughly the same penumbra.

9. You are in a dark room with a single incandescent 60-watt bulb in the center of the ceiling. You hold a book directly beneath the bulb and begin lowering it toward the floor. As the book is lowered, the size of the umbra
         a. gets bigger.
    *   b. gets smaller.
         c. stays the same size.

10. You are in a dark room with a 2-foot by 4-foot fluorescent light box in the center of the ceiling. You hold a book directly beneath the light box and begin lowering it toward the floor. As the book is lowered, the size of the umbra
    *   a. gets bigger.
         b. gets smaller.
         c. stays the same size.

11. What happens to the image produced by a pinhole camera when you move the back wall farther from the pinhole? It becomes
    *   a. larger and fainter.
         b. smaller and fainter.
         c. larger and brighter.
         d. smaller and brighter.

12. When the object moves farther from the pinhole, the image produced by the pinhole camera
         a. gets larger.
    *   b. gets smaller.
         c. remains the same size.

Physics: A World View, Sixth Edition by Larry Kirkpatrick and Gregory Francis

13. What effect does enlarging the hole in a pinhole camera have on the image? The image gets
      a. dimmer and fuzzier.
      b. brighter and sharper.
      c. dimmer and sharper.
  *   d. brighter and fuzzier.

14. What effect does enlarging the hole in a pinhole camera have on the image? The image gets
      a. larger.
      b. smaller.
      c. sharper.
  *   d. fuzzier.

15. A camera obscura used by a portrait painter is located 6 m from a child who stands 1 m tall. How tall is her image if the back of the camera obscura is 2 m away?
  *   a. 1/3 m
      b. 1/2 m
      c. 1 m
      d. 3 m

16. When a ray of light hits most surfaces, it
      a. reflects off in one specific direction.
      b. bounces off in the same direction.
      c. passes right through the surface.
  *   d. scatters off in many directions.

17. The law of reflection says that
      a. light travels in straight lines.
      b. light must be reflected into our eyes to be seen.
  *   c. the angle of reflection is equal to the angle of incidence.
      d. light striking a rough surface is scattered in many directions.

18. How long is the image of a meter stick produced by a flat mirror?
  *   a. one meter
      b. less than one meter
      c. more than one meter

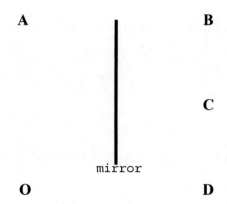

19. Which letter corresponds to the location of the image of the object O shown above?
    a. A
    b. B
    c. C
    * d. D

20. The shortest mirror in which a creature from outer space can see its entire body is _____ its height.
    a. twice
    b. equal to
    * c. one half
    d. The length of the mirror depends on how far away the creature stands.

21. In order for you to just be able to see the top of your head, the top of the mirror must be
    a. even with your chin.
    b. at eye level.
    * c. midway between your eyes and the top of your head.
    d. level with the top of your head.

22. Which, if any, direction does a flat mirror reverse?
    a. up and down
    b. left and right
    * c. front and back
    d. It does not reverse a direction.

23. If you stand 2 ft in front of a flat mirror, how far away from you is your image?
    a. 1 ft
    b. 2 ft
    c. 3 ft
    * d. 4 ft

24. If a 4-ft-tall child stands 2 ft in front of a vertical plane mirror, the image of the child will be _____ ft tall.
    a. 1
    b. 2
 * c. 4
    d. 8

25. How do the size and location of your image change as you walk away from a flat mirror? The image
    a. remains the same size and the same distance away.
    b. becomes smaller and moves further away.
    c. becomes smaller and remains the same distance away.
 * d. remains the same size and moves further away.

26. The images produced by opposing mirrors appear to get progressively smaller because they are progressively
    a. smaller.
 * b. farther away.
    c. smaller and farther away.
    d. fainter.

27. How many images are formed by two mirrors at 45° to each other?
    a. 1
    b. 3
    c. 5
 * d. 7

28. A ray reflected from a retroreflector
    a. has an angle of reflection equal to the angle of incidence.
    b. passes through the focal point.
    c. forms a right angle with in incident ray.
 * d. travels in the direction opposite that of the incident ray.

29. If rays of light parallel to the optic axis are incident on a _____ mirror, they converge to a point after leaving the mirror.
    a. flat
    b. convex
 * c. concave
    d. diffuse

30. If rays of light parallel to the optic axis are incident on a _____ mirror, they appear to diverge from a point after leaving the mirror.
    - a. flat
    - * b. convex
    - c. concave
    - d. diffuse

31. A ray of light parallel to the optic axis of a concave mirror is reflected back
    - a. through the center of curvature for the mirror.
    - * b. through the focal point.
    - c. parallel to the optic axis.
    - d. as if it came from the focal point.

32. A ray of light parallel to the optic axis of a convex mirror is reflected back
    - a. through the center of curvature for the mirror.
    - b. through the focal point.
    - c. parallel to the optic axis.
    - * d. as if it came from the focal point.

33. What type of mirror would you use to produce a magnified image of your face?
    - a. flat
    - * b. concave
    - c. convex
    - d. You could use a concave or a convex mirror.

34. The back surfaces of automobile headlights are curved
    - a. because inverted, real images of filaments shine brighter.
    - * b. to concentrate light in one direction.
    - c. for structural reasons not related to optics.
    - d. to get multiple images of the filament.

35. The focal length of a curved mirror is _____ the radius of the spherical surface.
    - a. twice
    - b. equal to
    - * c. half
    - d. one-quarter

36. What is the difference between a real image and a virtual image?
    - a. A real image is always erect; a virtual image can be inverted.
    - b. A real image is bigger than the object; a virtual image can be smaller.
    - * c. Light comes from a real image; it only appears to come from a virtual one.
    - d. There is no difference; what you get depends only on the type of mirror.

37. The image of a child standing in front of a flat mirror is
    * a. erect and virtual.
    b. erect and real.
    c. inverted and virtual.
    d. inverted and real.

38. The image formed by a bathroom mirror is
    * a. erect and virtual.
    b. erect and real.
    c. inverted and virtual.
    d. inverted and real.

39. A ray of light passing through the focal point at an angle to the optic axis of a concave mirror is reflected back
    a. through the center of the sphere.
    b. through the focal point.
    * c. parallel to the optic axis.
    d. in the horizontal direction.

40. A ray of light heading toward the focal point at an angle to the optic axis of a convex mirror is reflected back
    a. through the center of the sphere.
    b. through the focal point.
    * c. parallel to the optic axis.
    d. in the horizontal direction.

41. What type of image is formed when an object is placed midway between the focal point and the center of a concave mirror?
    a. erect and virtual
    b. inverted and virtual
    c. erect and real
    * d. inverted and real

42. What type of image is formed when an object is placed at a distance equal to 3 focal lengths from a concave mirror?
    a. erect and virtual
    b. inverted and virtual
    c. erect and real
    * d. inverted and real

43. What type of image is formed when an object is placed at a distance of 1.5 focal lengths from a convex mirror?
    * a. erect and virtual
    b. inverted and virtual
    c. erect and real
    d. inverted and real

44. A concave mirror has a focal length of 20 cm. Where is the image located when an object is placed 60 cm from the mirror?
    a. 15 cm in front
    * b. 30 cm in front
    c. 60 cm in front
    d. 15 cm in back

45. Where is the image located when an object is placed 60 cm from a convex mirror with a focal length of 20 cm?
    * a. 15 cm in back
    b. 30 cm in back
    c. 60 cm in back
    d. 15 cm in front

46. How far in front of a concave spherical mirror with a focal length f would you place a candle so that it appears to burn at both ends?
    a. $4f$
    * b. $2f$
    c. f
    d. $1/2 f$

47. The speed of light
    a. has never been measured.
    b. is about the same as that of sound.
    c. is infinitely fast.
    * d. is very fast, but not infinite.

48. Light from our nearest neighbor star takes 4.4 years to reach Earth. What can we say about what is happening on this star at this moment?
    * a. nothing
    b. the same thing that happened 4.4 years ago
    c. Its surface has gotten cooler.
    d. It has become a red giant.

49. If the Sun is 150 million km away from Earth, how long does it take Sunlight to reach Earth?
    a. 0.5 s
    b. 15 s
    c. 45 s
    * d. 500 s

50. If the Moon is 400,000 km away from Earth, how long does it take Moonlight to reach Earth?
    a. 0.75 s
    * b. 1.3 s
    c. 4 s
    d. 13 s

51. If Galileo and his assistant were 15 km apart, how long would it take light to make a round trip?
    a. 10^{-6} s
    b. 10^{-5} s
    * c. 10^{-4} s
    d. 10^{-3} s

52. When we say a book is green, we mean that it
    a. absorbs green light.
    b. reflects all colors equally.
    * c. reflects green light and absorbs the others.
    d. absorbs red light and reflects the others.

53. What color would you see if a white piece of paper is simultaneously illuminated by red and green light?
    a. white
    b. black
    c. magenta
    * d. yellow

54. What color would you see if a white piece of paper is simultaneously illuminated by red and blue light?
    a. white
    b. black
    * c. magenta
    d. yellow

55. What color would you expect to see if you remove the red light from white light?
      a. yellow
  * b. cyan
      c. magenta
      d. green

56. A substance is known to reflect red and blue light. What color would it have when it is illuminated by white light?
      a. yellow
      b. cyan
  * c. magenta
      d. white

57. A substance is known to reflect red and blue light. What color would it have when it is illuminated by green light?
      a. yellow
      b. cyan
      c. magenta
  * d. black

58. A CREST toothpaste tube viewed under white light has a red C on a white background. However, when we use red light, the C is not visible because the
  * a. background appears red.
      b. C appears white.
      c. C appears black.

59. In the discussion of the color properties of lights, complementary colors are ones that
      a. go well together.
  * b. add to form white.
      c. add to form black.
      d. add to form the third primary color.

60. What is the complementary color to magenta?
      a. red
  * b. green
      c. blue
      d. yellow

61. What is the complementary color to red?
    - a. green
    * b. cyan
    - c. magenta
    - d. yellow

62. What is the complementary color to blue?
    - a. green
    - b. cyan
    - c. magenta
    * d. yellow

63. What is the complementary color to green?
    - a. green
    - b. cyan
    * c. magenta
    - d. yellow

64. The sky appears blue because
    - a. that is its natural color.
    - b. Earth's atmosphere emits blue light.
    - c. the air away from the Sun cools down and turns blue.
    * d. Earth's atmosphere scatters more blue light than red.

65. The Sun appears yellow because
    - a. that is its natural color.
    - b. its surface emits more yellow light than any other color.
    - c. the Sunlight cools off as it passes through Earth's atmosphere.
    * d. Earth's atmosphere scatters more blue light than red.

66. If the atmosphere primarily scattered green light instead of blue light, the sky would appear ___ and the Sun would appear _____ .
    * a. green … magenta
    - b. green … yellow
    - c. magenta … green
    - d. magenta … yellow

Chapter 18: Refraction of Light

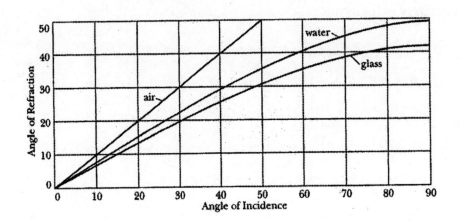

1. According to the graph given above, at what angle is light refracted if it strikes the surface of water at an angle of 30°?
    - a. 20°
    - * b. 23°
    - c. 42°
    - d. 48°

2. According to the graph given above, at what angle is light refracted if it strikes the surface of glass at an angle of 30°?
    - * a. 20°
    - b. 23°
    - c. 42°
    - d. 48°

3. Light in water is incident at the surface with air at an angle of 30°. According to the graph given above, at what angle is it refracted?
    - a. 20°
    - b. 23°
    - * c. 42°
    - d. 48°

4. Light in glass is incident at the surface with air at an angle of 30°. According to the graph given above, at what angle is it refracted?
    - a. 20°
    - b. 23°
    - c. 42°
    - * d. 48°

5. If a ray of light strikes a pane of glass at 45° to the normal, it
      a. passes straight through as if the glass were not there.
      b. leaves the glass at a smaller angle to the normal.
      c. leaves the glass at a larger angle to the normal.
   *   d. leaves with the same angle to the normal, but is deflected to the side.

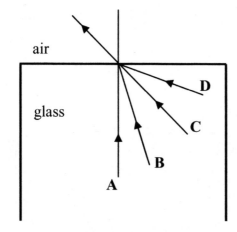

6. A narrow beam of light emerges from a block of glass in the direction shown above. Which arrow best represents the path of the beam within the glass?
      a. A
   *   b. B
      c. C
      d. D

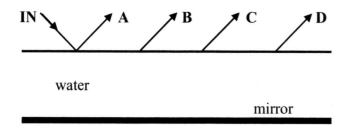

7. A mirror is lying on the bottom of a fish tank that is filled with water as shown above. If IN represents a light ray incident on the top of the tank, which possibility best represents the ray that comes out?
      a. A
   *   b. B
      c. C
      d. D

8. Two coins are at equal distances from your eye. One is under 40 cm of water, the other under 40 cm of glass. Which coin appears closer?
    *   a. the one under the glass
        b. the one under the water
        c. Neither, they both appear at the same distance.

9. You are standing at the edge of a swimming pool filled with water looking at a logo painted on the bottom. The logo appears to be _____ it actually is.
    *   a. closer to you than
        b. further from you than
        c. the same distance away as

10. You are lying on the bottom of a swimming pool filled with water looking at a balloon suspended over the water. The balloon appears to be _____ it actually is.
        a. closer to you than
    *   b. further from you than
        c. the same distance away as

11. Due to the refraction of light, fish in an aquarium appear to be _____ they really are.
    *   a. closer than
        b. further away than
        c. located where

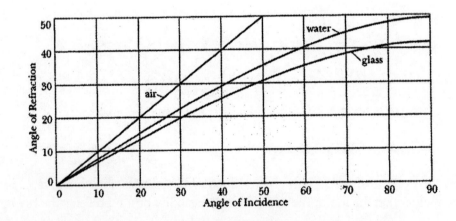

12. According to the graph shown above, what is the critical angle for a water-air surface?
        a. 42°
    *   b. 48°
        c. 50°
        d. 90°

13. According to the graph shown above, what is the critical angle for a glass-air surface?
    *   a. 42°
        b. 48°
        c. 50°
        d. 90°

14. The critical angle for total internal reflection at an air-water interface is approximately 48°. In which of the following situations will total internal reflection occur?
        a. light incident in water at 40°
    *   b. light incident in water at 55°
        c. light incident in air at 40°
        d. light incident in air at 55°

15. The telephone companies are proposing using "light pipes" to carry telephone signals between various locations. The light is contained inside these pipes
        a. because they are coated with silver.
        b. as long as they are straight.
    *   c. by total internal reflection if the curves are not too sharp.
        d. because laser light does not travel well through air.

16. There is a limit to how much a fiber-optic cable can be bent before light "leaks" out. This is because bending the pipe allows light to strike the surface at angles less than the critical angle. If you were laying fiber-optic cable under water instead of in air, this problem would be
    *   a. even greater.
        b. not as severe.
        c. no different.

17. While transmitting white light down a fiber-optic cable, you bend the cable too much in one place, and some of the light "leaks" out. Which is the first color of light to leak out?
    *   a. red
        b. green
        c. yellow
        d. blue

18. Because of the refraction of light in the atmosphere, the apparent position of a star is usually _____ its true position.
    *   a. higher than
        b. lower than
        c. the same as

19. If you were going to send a beam of light to the Moon when it is just above the horizon, you would aim
    a. high.
    b. low.
 * c. directly at the Moon.

20. Because of the refraction of light in the atmosphere, the appearance of the Sun or Moon changes as it approaches the horizon. This change can be described as
    a. an increase in diameter.
    b. a decrease in diameter.
 * c. a shortening in the vertical direction.
    d. a shortening in the horizontal direction.

21. Because of atmospheric refraction the image of the Sun rises _____ and sets _____ than the actual Sun.
    a. earlier ... earlier
 * b. earlier ... later
    c. later ... later
    d. later ... earlier

22. When two stars are observed toward the south over the equator, they are separated by 5° along an east-west arc. When these stars are later observed near the horizon, their separation will be
    a. more than 5°.
    b. the same.
 * c. less than 5°.

23. In the absence of an atmosphere, a star moves across the sky from horizon to horizon at a constant angular speed. In the presence of an atmosphere, the star will appear to move
    a. slower during the middle of its path.
    b. faster on rising and slower on setting.
 * c. slower on rising and setting.
    d. slower on rising and faster on setting.

24. Looming is observed when light is bent _____ near the ground.
    a. upward by warm air
 * b. downward by cold air
    c. upward by cold air
    d. downward by warm air

25. When looming occurs, an image is seen that is _____ the object.
    a. lower than
    * b. higher than
    c. at the same height as

26. Mirages are observed when light is bent _____ near the ground.
    * a. upward by warm air
    b. downward by cold air
    c. upward by cold air
    d. downward by warm air

27. When a mirage occurs, an image is seen that is _____ the object.
    * a. lower than
    b. higher than
    c. at the same height as

28. ROY G. BIV is
    a. a cowboy singer.
    b. a famous scientist who did much work in optics.
    * c. a mnemonic for remembering the colors in the rainbow.
    d. a way of remembering how mirages are formed.

29. Assume that the following colors of light pass through a prism. Which color ray is bent the most?
    a. red
    * b. blue
    c. green
    d. yellow

30. When white light passes through a prism, it is spread out into a rainbow of colors. This effect is known as
    a. interference.
    b. dispensation.
    c. complementarity.
    * d. dispersion.

31. The dispersion of light when it passes through a prism shows that
    a. the prism contains many narrow, equally spaced slits.
    b. all colors in the light are treated the same.
    * c. different colors have different indices of refraction.
    d. the speed of light in a vacuum is a constant.

32. Rainbows are due to
    a. reflection from the surface of raindrops.
    * b. refraction and reflection by raindrops.
    c. refraction through raindrops.
    d. refraction and reflection by ice crystals.

33. Which color is on the outside of the primary rainbow?
    * a. red
    b. yellow
    c. green
    d. violet

34. The order of the colors in the secondary rainbow is _____ those in the primary one.
    a. the same as
    * b. the reverse of
    c. completely different than

35. Where in the sky would you expect to see a rainbow in the evening?
    a. northern sky
    b. southern sky
    * c. eastern sky
    d. western sky

36. In order to see a rainbow at noon in the summer, a person must be looking
    a. east
    b. west
    c. north
    * d. down

37. At what time of day would you expect to see the top of a rainbow rise?
    a. Sunrise
    b. mid morning
    * c. mid afternoon
    d. Sunset

38. At what time of day would you expect to see the top of a rainbow set?
    a. Sunrise
    * b. mid morning
    c. mid afternoon
    d. Sunset

39. You see a rainbow from the window of your plane at noon. Where is the rainbow?
    a. in the east
    b. in the west
    c. in front of the plane
    * d. below the plane

40. Where in the sky would you look to see the 22° halo?
    a. 22° above the horizon
    * b. 22° from the Sun in any direction
    c. 22° above the Sun
    d. 22° below the Sun

41. Where in the sky would you look to see Sundogs?
    a. 22° above the Sun
    b. 22° below the Sun
    * c. 22° to the left and right of the Sun
    d. 22° to the left and right of the antisolar position

42. The halos that we sometimes see around the Sun and Moon are due to
    a. reflection in raindrops.
    b. refraction in raindrops.
    c. reflection from ice crystals.
    * d. refraction in ice crystals.

43. If rays of light parallel to the optic axis are incident on a _____ lens, they converge to a point after leaving the lens.
    a. flat
    * b. converging
    c. diverging
    d. cylindrical

44. A lens which is thinner at the center than at the edges is a _____ lens.
    a. converging
    * b. diverging
    c. converging or diverging
    d. coherent

45. A lens which is thicker at the center than at the edges is a _____ lens.
    * a. converging
    b. diverging
    c. converging or diverging
    d. coherent

46. For a converging lens, a ray arriving parallel to the optic axis
    a. appears to come from the principal focal point.
    * b. passes through the principal focal point.
    c. passes through the "other" focal point.
    d. appears to come from the "other" focal point.

47. For a diverging lens, a ray arriving parallel to the optic axis
    a. heads toward the principal focal point.
    * b. appears to come from the principal focal point.
    c. appears to come from the "other" focal point.
    d. heads toward the "other" focal point.

48. Two lenses with identical shapes are made from glasses with different indices of refraction. Which one has the shorter focal length?
    * a. the one with the higher index of refraction
    b. the one with the lower index of refraction
    c. The focal lengths are the same.

49. What type of lens would you use to construct a slide projector?
    * a. converging
    b. diverging
    c. coherent
    d. diverse

50. For a diverging lens, a ray that leaves parallel to the optic axis
    a. was heading toward the principal focal point.
    b. passes through the principal focal point.
    c. passes through the "other" focal point.
    * d. was heading toward the "other" focal point.

51. The focal length of a converging lens is 20 cm. The image of an object placed 60 cm from the center of this lens is
    * a. real and inverted.
    b. real and erect.
    c. virtual and inverted.
    d. virtual and erect.

52. The focal length of a diverging lens is 20 cm. The image of an object placed 40 cm from the center of this lens is
      a. real and inverted.
      b. real and erect.
      c. virtual and inverted.
*   d. virtual and erect.

53. A converging lens is used to form a sharp image of a candle. If the lower half of the lens is covered by a piece of paper, the
      a. lower half of the image will disappear.
      b. upper half of the image will disappear.
*   c. image will become dimmer.
      d. image will not change.

54. A converging lens is used to form a sharp image of a candle. If the lens is covered by a piece of paper with a small hole in the center, the
      a. outer part of the image will disappear.
      b. inner part of the image will disappear.
*   c. image will become dimmer.
      d. image will not change.

55. A converging lens has a focal length of 20 cm. Where is the image located when an object is placed 60 cm from the lens?
      a. 15 cm on the far side
*   b. 30 cm on the far side
      c. 60 cm on the far side
      d. 15 cm on the near side

56. A diverging lens has a focal length of 20 cm. Where is the image located when an object is placed 60 cm from the lens?
*   a. 15 cm on the near side
      b. 30 cm on the near side
      c. 60 cm on the near side
      d. 15 cm on the far side

57. A camera employs a _____ lens to form _____ images.
*   a. converging .... real
      b. converging .... virtual
      c. diverging .... real
      d. diverging .... virtual

58. The image formed on the film of a camera is
      a. virtual and erect.
      b. real and erect.
      c. virtual and inverted.
  * d. real and inverted.

59. In most cameras the location of the image is adjusted to appear on the film by changing the
  * a. position of the lens.
      b. diameter of the diaphragm.
      c. shape of the lens.
      d. focal length of the lens.

60. What kind of lens would you place in front of the lens of a simple camera to turn it into a close-up camera for taking pictures of small objects?
  * a. converging
      b. diverging
      c. cylindrical
      d. astigmatic

61. Which of the following CANNOT be used to reduce spherical aberration?
      a. use a diaphragm to reduce the effective diameter of the lens
      b. grind the lens with parabolic surfaces
      c. use a combination of lenses rather than a single lens
  * d. use a glass with a higher index of refraction

62. What is the purpose of the diaphragm in a camera?
      a. to adjust the focal length of the lens
      b. to turn the image right side up
  * c. to adjust the amount of light passing through the lens
      d. to change the width of the field of view

63. Which of the following can be used most effectively to reduce chromatic aberration?
      a. use a diaphragm to reduce the effective diameter of the lens
      b. grind the lens with parabolic surfaces
  * c. use a combination of lenses rather than a single lens
      d. use a more sensitive color film

64. A human eye employs a _____ lens to form _____ images.
  * a. converging .... real
      b. converging .... virtual
      c. diverging .... real
      d. diverging .... virtual

65. The image formed on the retina of an eye is
    a. virtual and erect.
    b. real and erect.
    c. virtual and inverted.
    * d. real and inverted.

66. What is the purpose of the pupil in an eye?
    a. to adjust the focal length of the lens
    * b. to adjust the amount of light passing through the lens
    c. to focus the image
    d. to change the width of the field of view

67. In the human eye the location of the image is adjusted to appear on the retina by changing the
    a. position of the lens.
    b. diameter of the pupil.
    c. shape of the cornea.
    * d. focal length of the lens.

68. Most of the refractive power in the eye is due to the
    a. lens.
    b. retina.
    * c. cornea.
    d. iris.

69. Astigmatism is caused by
    a. the dispersion of light as it passes through the lens.
    * b. the non-spherical shape of the cornea or lens.
    c. a non-circular shape of the pupil.
    d. weak nerve endings in the retina.

70. A converging lens that has a focal length of 0.25 m has a value of _____ diopters.
    a. +0.25
    * b. +4
    c. -0.25
    d. -4

71. A +5-diopter lens is a _____ lens with a focal length of _____.
    a. converging ... 5 m
    b. diverging ... 5 m
    * c. converging ... 0.2 m
    d. diverging ... 0.2 m

72. How many diopters are there for a converging lens with a focal length of 0.4 m?
     a. –2.5
     b. –0.4
     c. +0.4
*   d. +2.5

73. How many diopters are there for a diverging lens with a focal length of 0.4 m?
*   a. –2.5
     b. –0.4
     c. +0.4
     d. +2.5

74. What focal-length lens would you need to place next to a converging lens of focal length 25 cm to create an effective focal length of 20 cm for the combination?
     a. – 100 cm
     b. – 5 cm
     c. + 5 cm
*   d. + 100 cm

75. A converging lens of focal length 50 cm is placed next to a diverging lens of length 25 cm. What is the effective focal length for this combination?
*   a. –50 cm
     b. – 25 cm
     c. + 25 cm
     d. + 50 cm

76. How far from a magnifying glass do you place an object to view it?
*   a. closer than the focal length
     b. a distance equal to the focal length
     c. greater than the focal length but less than twice the focal length
     d. at twice the focal length

77. Telescopes that use mirrors as the objectives are known as
     a. refractors.
*   b. reflectors.
     c. Galilean telescopes.
     d. terrestrial telescopes.

78. A company wants to build a telescope with an objective that is 1 meter in diameter. Which of the following is NOT an advantage of a reflector over a refractor?
     a. The mirror can be supported from behind.
     b. The mirror does not have chromatic aberration.
     c. The mirror does not sag under its own weight.
 *   d. The mirror gathers a lot more light.

79. How does the length of a refracting telescope compare to the focal lengths of the lenses? It is equal to _____ of the two focal lengths.
     a. the differences
 *   b. the sum
     c. one-half the sum
     d. twice the sum

Chapter 19: A Model for Light

1. Newton's idea that a light beam consists of tiny particles works for the law of reflection provided that the surface is
   - a. frictionless.
   - b. perfectly elastic.
   - * c. frictionless and perfectly elastic.
   - d. transparent.

2. Newton's idea that a light beam consists of tiny particles could successfully explain the law of refraction. His idea included
   - a. a force acting at the surface.
   - b. a force acting normal to the surface.
   - c. a force directed into the medium with the higher index of refraction.
   - * d. All of the above features are correct.

3. When a wave crosses a boundary (for example, water waves going from deep to shallow water), the property that remains <u>unchanged</u> is the
   - a. wavelength.
   - b. velocity.
   - * c. frequency.
   - d. amplitude.

4. The property of a light wave that determines its brightness is its
   - a. wavelength.
   - b. frequency.
   - c. speed.
   - * d. amplitude.

5. Newton's idea that light consists of tiny particles predicted that the speed of light would be _____ in water than in air.
   - a. slower
   - * b. faster
   - c. the same

6. Newton's idea that a light beam consists of tiny particles did NOT correctly predict
   - a. that light beams travel in straight lines.
   - b. that the angle of incidence equals the angle of reflection.
   - c. the law of refraction.
   - * d. that the speed of light in substances with high refractive indices is lower than that in air.

7. If particles incident at 45° from the normal strike a completely elastic surface that has friction, the angle of reflection (with respect to the normal) will be
    a. greater than 45°.
    b. equal to 45°.
*   c. less than 45°.

8. If particles incident at 45° from the normal strike a frictionless surface that is not completely elastic, the angle of reflection (with respect to the normal) will be
*   a. greater than 45°.
    b. equal to 45°.
    c. less than 45°.

9. Measurements show that the speed of light in glass is _____ than in air as predicted by the _____ theory of light.
    a. slower ... particle
*   b. slower ... wave
    c. faster ... particle
    d. faster ... wave

10. The dispersion of light when it passes through a prism shows that
    a. the prism contains many narrow, equally spaced slits.
    b. all wavelengths have the same speed in a material.
*   c. different wavelengths have different speeds in the material.
    d. the index of refraction is the same for all wavelengths.

11. The dispersion of light when it passes through a prism shows that
    a. the prism contains many narrow, equally spaced slits.
    b. all wavelengths have the same speed in a material.
    c. the index of refraction is the same for all wavelengths.
*   d. red light has a larger speed in a material than blue light.

12. If the speed of light in a vacuum is 300 million m/s and the index of refraction for light in a piece of glass is 1.5, the speed of light in this glass is _____ million m/s.
    a. 150
*   b. 200
    c. 300
    d. 450

13. The speed of light in diamond is 1.24×10^8 m/s. What is the index of refraction for diamond?
      a. 0.41
      b. 1
      c. 1.24
  * d. 2.42

14. The speed of light in glass is approximately _____ that in air.
      a. 100 times faster than
      b. 100 times slower than
      c. 50% faster than
  * d. 33% slower than

15. The speed of light in water is approximately _____ that in air.
      a. the same as
      b. 100% faster than
      c. 33% faster than
  * d. 25% slower than

16. If it takes light 5 ns to travel 1 m in an optical cable, what is the index of refraction of the cable?
      a. 0.67
  * b. 1.5
      c. 2
      d. 5

17. If an optical cable has an index of refraction of 1.5, how long will it take a signal to travel between two points on opposite coasts of the United States separated by a distance of 5000 km?
      a. 2.5×10^{-5} s
      b. 1.1×10^{-2} s
      c. 1.7×10^{-2} s
  * d. 2.5×10^{-2} s

18. Red light is used to form a two-slit interference pattern on a screen. As the two slits are moved further apart, the separation of the bright bands on the screen
  * a. decreases.
      b. increases.
      c. remains the same.

19. The observation that the two-slit interference pattern produced by blue light is narrower than that produced by red light indicates that blue light
    a. has a longer wavelength.
    * b. has a shorter wavelength.
    c. travels faster.
    d. has a lower frequency.

20. A closely spaced pair of narrow slits is illuminated by red light (wavelength = 700 nm) and then by blue light (wavelength = 450 nm). If the pattern produced by the slits is projected on a screen at a large distance from the slits, which color light will produce the largest spacing?
    * a. red
    b. blue
    c. Neither; they are the same.

21. Which of the following colors produces the narrowest interference pattern?
    a. orange
    b. green
    * c. violet
    d. red

22. If red light has a wavelength of 600 nm, its frequency is
    * a. 5×10^{14} Hz.
    b. 1.8×10^{11} Hz.
    c. 2×10^{-14} Hz.
    d. 5×10^{7} Hz.

23. What is the wavelength of light that has a frequency of 5×10^{14} Hz?
    * a. 6×10^{-7} m
    b. 6×10^{-10} m
    c. 1.5×10^{23} m
    d. 2×10^{-6} m

24. Holograms may be viewed with
    a. light from a laser.
    b. sunlight.
    c. a spot light.
    * d. all of the above.

25. Which of the following is NOT true of a holographic puzzle?
       a. Each piece of the holographic puzzle sees the scene from a different perspective.
    * b. A holographic puzzle can only be viewed with laser light.
       c. A holographic puzzle can be made on a thin piece of film.
       d. Holographic puzzles of three-dimensional objects must be made with laser light.

26. What happens to the size of the image produced by a hologram that was made with red light when it is illuminated by blue light? The image
       a. is larger.
    * b. is smaller.
       c. remains the same size.

27. Red light is used to form a single-slit diffraction pattern on a screen. As the slit is made narrower, the separation of the bright bands on the screen
       a. decreases.
    * b. increases.
       c. remains the same.

28. The observation that the single-slit diffraction pattern produced by red light is wider than that produced by blue light indicates that red light
    * a. has a longer wavelength.
       b. has a shorter wavelength.
       d. has a higher index of refraction.
       e. has a higher frequency.

29. The widest single-slit diffraction pattern would be caused by light of wavelength _____ passing through a slit of width _____ .
    * a. 800 nanometer … 500 nanometer
       b. 800 nanometer … 400 nanometer
       c. 450 nanometer … 400 nanometer
       d. 600 nanometer … 500 nanometer

30. Which of the following colors produces the widest diffraction pattern?
       a. orange
       b. green
       c. violet
    * d. red

31. Which of the following phenomena does NOT show a difference between the wave theory and particle theory of light?
    *   a. reflection
         b. refraction
         c. interference
         d. diffraction

32. Newton's idea that a light beam consists of tiny particles did NOT correctly predict
         a. that light beams travel in straight lines.
         b. the law of reflection.
         c. the law of refraction.
    *   d. the existence of diffraction and interference effects.

33. The colors seen in soap bubbles result from
         a. diffraction.
         b. dispersion.
    *   c. interference.
         d. soap pigmentation.

34. If light with a wavelength of 600 nm in air passes through a pane of glass with an index of refraction of 1.5, the wavelength of the light in the glass is
    *   a. 400 nm.
         b. 600 nm.
         c. 900 nm.
         d. 1200 nm.

35. If light with a frequency of 6×10^{14} Hz in air passes through a pane of glass with an index of refraction of 1.5, the frequency of the light in the glass is
         a. 4×10^{14} Hz.
    *   b. 6×10^{14} Hz.
         c. 9×10^{14} Hz.
         d. 1.2×10^{15} Hz.

36. Light from a sodium lamp with a wavelength in a vacuum of 590 nm enters diamond in which the speed of light is 1.24×10^8 m/s. What is the wavelength of this light in diamond?
    *   a. 244 nm
         b. 476 nm
         c. 732 nm
         d. 1430 nm

37. A thin soap film in air will strongly reflect light when its thickness is _____ the wavelength of the light in the film.
    a. 1/8 of
    * b. 1/4 of
    c. 1/2 of
    d. equal to

38. A thin soap film in air will strongly transmit light when its thickness is _____ the wavelength of the light in the film.
    a. 1/8 of
    b. 1/4 of
    c. 1/3 of
    * d. 1/2 of

39. A thin film of air between two panes of glass will strongly reflect light when its thickness is _____ the wavelength of the light.
    a. 1/8 of
    * b. 1/4 of
    c. 1/2 of
    d. equal to

40. If a film used to coat a photographic lens has an index of refraction smaller than that of the lens, light will be strongly reflected when the film thickness is _____ the wavelength of the light in the film.
    a. 1/8 of
    b. 1/4 of
    * c. 1/2 of
    d. equal to

41. What is the thinnest soap film that will strongly reflect light with a wavelength of 400 nm in the film?
    a. 50 nm
    * b. 100 nm
    c. 200 nm
    d. 400 nm

42. What is the thinnest soap film that will strongly reflect red light from a helium–neon laser? The wavelength of this light is 633 nm in air and 470 nm in soapy water.
    * a. 118 nm
    b. 158 nm
    c. 235 nm
    d. 317 nm

Physics: A World View, Sixth Edition by Larry Kirkpatrick and Gregory Francis

43. Which of the following effects occurs for transverse waves but not for longitudinal waves?
    a. interference
    b. diffraction
  * c. polarization
    d. refraction

44. We know that light is a transverse rather than a longitudinal wave, because the _____ of a longitudinal wave is meaningless.
    a. refraction
    b. interference
    c. diffraction
  * d. polarization

45. If pieces of cellophane are placed between two Polaroids, we see multicolored patterns. These colors are due to
    a. thin film interference.
    b. dispersion.
    c. the dependence of speed on wavelength.
  * d. the dependence of the rotation of the plane of polarization on wavelength.

46. If all the labels have come off the sunglasses in the drug store, you could tell which ones were polarized by
    a. feeling their surfaces.
    b. scratching the lenses to see if they are plastic.
  * c. looking at sunlight reflected from water.
    d. measuring the focal lengths of the lenses.

47. If all the labels have come off the sunglasses in the drug store, you could tell which ones were polarized by
    a. feeling their surfaces.
    b. scratching the lenses to see if they are plastic.
  * c. looking through two lenses and rotating one.
    d. looking at the light from florescent lights.

48. Polarized sunglasses
    a. absorb both the horizontal and vertical polarizations.
  * b. absorb the horizontal polarization and transmit the vertical polarization.
    c. rotate both polarizations.
    d. absorb the vertical polarization and transmit the horizontal polarization.

Chapter 20: Electricity

1. Which of the following can be used to charge a metal rod that is held in your hand? Rub it with
    a. fur.
    b. silk.
    c. wool.
    * d. It can't be done.

2. A metal rod held in your hand cannot be charged by rubbing it with cloth, fur, or plastic, or by contact with another charged object. From this we conclude that
    a. the metal rod is a conductor.
    b. your body is a conductor.
    * c. both the metal rod and your body are conductors.

3. A glass rod held in your hand can be charged by rubbing it with silk or a plastic bag. From this observation you can conclude that glass is
    a. a conductor.
    * b. an insulator.
    c. Nothing can be concluded.

4. A neutral metal sphere is hanging from the ceiling by a thin insulating thread. Which of two equally charged rods, one metal and one rubber, would transfer the greater amount of charge to the metal sphere when touched to the sphere?
    * a. the metal rod
    b. the rubber rod
    c. Both rods would transfer the same charge.
    d. There is not enough information to say.

5. Which of the following is a conductor?
    a. plastic
    * b. human body
    c. glass
    d. silk

6. Which of the following is NOT a conductor?
    a. human body
    b. moisture
    c. silver
    * d. plastic

7. We are led to believe that there were two kinds of charge because we are able to
     a. isolate them in separate containers.
   * b. see two effects, attraction and repulsion.
     c. detect different weights for the same object.
     d. separate electrons and protons.

8. If two balloons are each rubbed with wool, we find that the balloons
     a. attract each other.
     b. do not effect each other.
   * c. repel each other.

9. When silk is used to charge a glass rod, the glass obtains a net positive charge. Therefore we know that the silk has _____ charge.
     a. a net positive
   * b. a net negative
     c. no net

10. In the modern view of electricity, an object with a positive charge has
     a. an excess of positive charge.
     b. an excess of negative charge.
     c. a deficiency of negative charge.
   * d. an excess of positive charge or a deficiency of negative charge.

11. When a black plastic rod is rubbed by fur, it acquires a net negative charge of 10^{-5} C. Therefore, we can conclude that the fur
     a. acquired a net negative charge significantly less than 10^{-5} C.
     b. also acquired a net negative charge of 10^{-5} C.
     c. acquired a net positive charge significantly less than 10^{-5} C.
   * d. acquired a net positive charge of 10^{-5} C.

12. You have three identical metal spheres on insulating stands. The spheres hold charges $Q_A = -2q$, $Q_B = -q$, and $Q_C = 4q$. First, sphere A is brought into contact with sphere C and separated. Second, sphere C is brought into contact with sphere B and separated. What is the resulting charge on sphere B?
   * a. zero
     b. $+q/2$
     c. $+3q/2$
     d. $+5q/4$

13. A positively charged object repels
      a. negatively charged objects.
*   b. positively charged objects.
      c. neutral objects.

14. Which of the following objects are attracted by a positively charged object?
      a. only positively charged objects
      b. only negatively charged objects
      c. only neutral objects
*   d. neutral and negatively charged objects

15. A charged rod held close to an uncharged conducting object will
      a. repel the object.
*   b. attract the object.
      c. not effect the object because the object is uncharged.
      d. attract the object if the rod has a positive charge and repel it if the rod has a negative charge.

16. Neutral objects are attracted to both negatively and positively charged objects because
      a. they actually have a slight imbalance of charge.
      b. charge is conserved.
      c. they give off some of their charges to ground.
*   d. there is a redistribution of their internal charges.

17. A green rod is suspended by a thread so that it can rotate freely. When we bring a neutral rod near the green rod, it is attracted. This demonstrates that the green rod
      a. has a net positive charge.
      b. has a net negative charge.
      c. has equal numbers of positive and negative charges.
*   d. is charged, but not what sign the charge is.

18. You have three small balls, each hanging from an insulating thread. You find that balls 1 and 2 repel one another and that balls 2 and 3 repel one another. Balls 1 and 3 will
      a. attract each other.
*   b. repel each other.
      c. neither attract or repel each other.

19. You have three small balls, each hanging from an insulating thread. You find that balls 1 and 2 attract one another and that balls 2 and 3 repel one another. Which ball, if any, is possibly neutral?
    * a. ball 1
    b. ball 2
    c. ball 3
    d. None of the balls can be neutral.

20. A freshly wiped phonograph record attracts dust because
    * a. it has a charge from the rubbing that attracts the neutral dust.
    b. it has a negative charge that attracts the positive dust.
    c. it has a positive charge that attracts the negative dust.
    d. a neutral record attracts pieces of charged dust.

21. A rubbed balloon sticks to a wall because
    a. it has a negative charge that attracts the positive wall.
    b. it has a positive charge that attracts the negative wall.
    c. a neutral balloon attracts a charged wall.
    * d. it has a charge from the rubbing that attracts the neutral wall.

22. Which of the following sets of hypothetical observations about an object would demonstrate the existence of a third kind of electric charge?
    a. It repels positive and attracts negative charges.
    b. It attracts positive and repels negative charges.
    c. It attracts positive and negative charges.
    * d. It attracts positive and negative charges and repels itself.

23. An electroscope initially has a net negative charge. The foils come together when the electroscope is touched by a human hand because
    a. humans have a net positive charge.
    b. humans have a deficiency of negative charges.
    * c. the human body is a conductor.
    d. the human body is an insulator.

24. Two identical electroscopes, one initially charged and the other initially neutral, are connected by a thin rod. If both electroscopes are now charged, you can conclude that
    * a. they have the same charge.
    b. they have opposite charges.
    c. the rod is an insulator.
    d. one electroscope is grounded.

25. Two identical electroscopes, one initially charged and the other initially neutral, are connected by a thin rod. If both electroscopes are now charged, you can conclude that
    * a. the rod is a conductor.
      b. they have opposite charges.
      c. the rod is an insulator.
      d. one electroscope is grounded.

26. Suppose you find an electroscope with separated foils. When you bring a negatively charged rod slowly near it, the foils move further apart. You can then conclude that the electroscope was initially
      a. positively charged.
    * b. negatively charged.
      c. neutral.
      d. You cannot conclude anything about the initial charge.

27. Suppose you find an electroscope with separated foils. When you bring a positively charged rod slowly near it, the foils come together. You can then conclude that the electroscope was initially
      a. positively charged.
    * b. negatively charged.
      c. neutral.
      d. You cannot conclude anything about the initial charge.

28. An electroscope is initially neutral. Which of the following will allow you to use a positively charged rod to put a net negative charge on the electroscope?
      a. Touch the rod to the electroscope.
      b. Touch the rod to the electroscope and then momentarily touch it with your finger.
      c. Bring the rod near the electroscope, take the rod away, and then momentarily touch the electroscope with your hand.
    * d. Bring the rod near the electroscope, momentarily touch the electroscope with your hand, and then take the rod away.

29. An electroscope is initially neutral. Which of the following will allow you to use a positively charged rod to put a net positive charge on the electroscope?
    * a. Touch the rod to the electroscope.
      b. Touch the rod to the electroscope and then momentarily touch it with your finger.
      c. Bring the rod near the electroscope, take the rod away, and then momentarily touch the electroscope with your hand.
      d. Bring the rod near the electroscope, momentarily touch the electroscope with your hand, and then take the rod away.

30. Two charged objects are very, very far from any other charges. If the charge on one of them is changed in sign, the electric force between them
    a. doubles.
    b. is cut in half.
    c. stays the same.
    * d. reverses direction.

31. If the signs of the charges on two charged objects are both reversed, the electric force between them
    a. increases in size.
    b. decreases in size.
    c. changes direction.
    * d. does not change.

32. You have positively charged objects. Object B has twice the charge of object A. When they are brought close together, object A experiences an electric force of 10 newtons. The magnitude of the electric force experienced by object B is _____10 newtons?
    a. less than
    * b. equal to
    c. greater than

33. The electric force on a negative charge located near an isolated positive charge
    a. points away from the isolated charge.
    b. is tangent to a circle centered about the isolated charge.
    * c. points toward the isolated charge.
    d. is zero.

34. Two parallel plates have equal but opposite charges. The electric force on a negative charge placed near the center of the plates is
    * a. perpendicular to the plates and points toward the positive one.
    b. perpendicular to the plates and points toward the negative one.
    c. parallel to the plates.
    d. zero.

35. The electric charge on an electron _____ that on a proton.
    a. is identical to
    b. is larger than
    c. is smaller than
    * d. has the same size but the opposite sign as

36. Two charged objects are very, very far from any other charges. If the distance between them is doubled, the electric force between them
      a. doubles.
      b. is cut in half.
*   c. is 1/4th as large.
      d. quadruples.

37. Two charged objects are very, very far from any other charges. If the distance between them is cut in half, the electric force between them
      a. doubles.
      b. is cut in half.
      c. is 1/4th as large.
*   d. quadruples.

38. Two charged objects are very, very far from any other charges. If the charge on one of them is doubled while the other stays the same, the electric force between them
*   a. doubles.
      b. is cut in half.
      c. stays the same.
      d. quadruples.

39. Two charged objects are very, very far from any other charges. If the charge on one of them is doubled while the other is cut in half, the electric force between them
      a. doubles.
      b. is cut in half.
*   c. stays the same.
      d. quadruples.

40. Even though electric forces are very much stronger than gravitational forces, gravitational forces determine the motions in the Solar System because
      a. electric forces have a longer range than gravitational ones.
      b. electric forces have a shorter range than gravitational ones.
      c. the electric forces do not act through a vacuum.
*   d. the electric forces cancel due to the two kinds of charge.

41. The ratio of the electric and gravitational forces between an electron and a proton is _____ for different separations because
      a. different ... the charge-to-mass ratios are different.
      b. different ... the particles have different masses.
      c. the same ... the charges are the same.
*   d. the same ... both forces vary as the inverse square of the separation.

42. The ratio of the electric and gravitational forces between two protons is the same for all separations because
    * a. both forces vary as the inverse square of the separation.
    b. the charge-to-mass ratios are the same.
    c. the particles have the same masses.
    d. the charges are the same.

43. The force that holds atoms together is
    a. magnetic.
    b. gravitational.
    * c. electric.
    d. unknown to scientists.

44. It is not possible for two electric field lines to cross because _____ at that point.
    a. two electric charges cannot contribute to the field
    b. the field would have to be infinite
    * c. the field cannot have two different directions
    d. There is no problem with them crossing.

45. The value of the electric field at each point in space is defined to be the _____ charge at that point.
    a. energy of a unit positive
    b. energy of a unit negative
    * c. force on a unit positive
    d. force on a unit negative

46. What is the size of the electric field at a distance of 3 m from 2 C of positive charge if k is the constant in Coulomb's law?
    a. $k(1 \text{ C})(2 \text{ C})/(3 \text{ m})$
    b. $k(2 \text{ C})/(3 \text{ m})$
    c. $k(1 \text{ C})(2 \text{ C})/(3 \text{ m})^2$
    * d. $k(2 \text{ C})/(3 \text{ m})^2$

47. The electric field around an isolated positive charge
    * a. points outward.
    b. forms clockwise circles about the charge.
    c. points inward.
    d. forms counterclockwise circles about the charge.

48. The electric field around an isolated negative charge
    a. points outward.
    b. forms clockwise circles about the charge.
    * c. points inward.
    d. forms counterclockwise circles about the charge.

49. A −5-mC charge experiences a force of 2 N directed north. What is the electric field (magnitude and direction) at the location of the charge?
    a. 0.01 N/C directed north
    b. 0.01 N/C directed south
    c. 400 N/C directed north
    * d. 400 N/C directed south

50. A 15-mC charge experiences a force of 30 N directed west. What is the electric field (magnitude and direction) at the location of the charge?
    a. 0.45 N/C directed west
    b. 0.45 N/C directed east
    * c. 2000 N/C directed west
    d. 2000 N/C directed east

51. A metal sphere with a charge of +4 C experiences an electric force of 50 N directed to the left. If the charge on the sphere is changed to −8 C, what force will it experience?
    a. 50 N to the left
    b. 50 N to the right
    c. 100 N to the left
    * d. 100 N to the right

52. Two parallel plates have equal but opposite charges. The electric field lines near the center of the plates
    a. are perpendicular to the plates and point toward the positive plate.
    * b. are perpendicular to the plates and point toward the negative plate.
    c. are parallel to the plates.
    d. do not exist.

53. The numerical value of the electric potential energy at each point in space is defined to be the
    * a. work required to bring a charged object from the zero point.
    b. average force required to bring a charged object from the zero point.
    c. energy required to bring a unit of positive charge from the zero point.
    d. work required to bring a unit positive charge from the zero point.

54. The numerical value of the electric potential at each point in space is defined to be the
    a. work required to bring a charged object from the zero point.
    b. average force required to bring a charged object from the zero point.
    c. energy required to bring a charged object from the zero point.
    * d. work required to bring a unit positive charge from the zero point.

55. Points A and B have electric potentials of 12 V and 20 V, respectively. How much work would be required to take 3 C of positive charge from A to B?
    a. zero
    * b. 24 J
    c. 36 J
    d. 60 J

56. Points A and B each have an electric potential of +12 V. How much work would be required to take 3 C of positive charge from A to B?
    * a. zero
    b. 3 J
    c. 9 J
    d. 36 J

57. The electric potential energy of an object at point A is known to be 50 J. If it is released from rest at A, it gains 30 J of kinetic energy as it moves to point B. What is its potential energy at B?
    * a. 20 J
    b. 30 J
    c. 50 J
    d. 80 J

58. The electric potential energy of an object at point A is known to be 50 J. If it is released from rest at A, it gains 30 J of kinetic energy as it moves to point B. If the object has a charge of +2 C, what is the potential difference between points A and B?
    * a. 10 V
    b. 15 V
    c. 20 V
    d. 40 V

59. How much work does a 12-V battery do in pushing 2 mC of charge through a circuit containing one light bulb?
      a. 0.006 J
      b. 0.012 J
  * c. 0.024 J
      d. The answer depends on the type of bulb in the circuit.

60. What was your electric potential if a spark jumped 2 cm from your finger to a metal pipe?
      a. 30 V
      b. 60 V
      c. 30 kV
  * d. 60 kV

Chapter 21: Electric Current

1. Galvani and Volta performed experiments with a frog's leg. They discovered that the leg
   a. twitched when touched with an insulator.
   * b. twitched when touched with a metal different from the clamp holding it.
   c. twitched when touched with the same metal as the clamp holding it.
   d. had a net electric charge.

2. Which of the following statements about electric cells is NOT correct?
   a. Voltages add when cells are placed end to end.
   * b. The voltage of a cell depends on its size.
   c. Voltages of cells placed side by side remain the same.
   d. The voltage of a cell depends on the materials used.

3. When two identical batteries are connected in series, the total voltage of the combination is
   * a. the sum of the two voltages.
   b. equal to the voltage of one battery.
   c. the reciprocal of the sum of the reciprocals of the two voltages.

4. When two identical batteries are connected in parallel, the total voltage of the combination is
   a. the sum of the two voltages.
   * b. equal to the voltage of one battery.
   c. the reciprocal of the sum of the reciprocals of the two voltages

5. When two 6-V batteries are connected in parallel, the voltage across the combination is
   a. 3 V.
   * b. 6 V.
   c. 9 V.
   d. 12 V.

6. When two 6-V batteries are connected in series, the voltage across the combination is
   a. 3 V.
   b. 6 V.
   c. 9 V.
   * d. 12 V.

7. A flashlight has two D-cells (1.5 V) placed end to end. What voltage rating should the bulb have?
   a. 1.5 V
   b. 1.5 A
   * c. 3 V
   d. 3 A

8. Household electricity in the U.S.A. is
    a. 12-volt direct current.
    b. 12-volt alternating current.
    c. 120-volt direct current.
    * d. 120-volt alternating current.

9. Batteries provide _____ current, while your local power company provides _____ current.
    * a. direct ... alternating
    b. alternating ... alternating
    c. direct ... direct
    d. alternating ... direct

10. Combining the concept of a complete circuit with the conservation of charge leads to the conclusion that
    a. the amount of charge used up in a complete circuit depends on the type of bulbs.
    * b. whatever charge flows out of one end of the battery flows into the other end.
    c. two bulbs in parallel use up twice as much charge as a single bulb.
    d. each bulb in a series uses up one-half of the charge.

11. Which of the following is true for a filament in a bulb connected to a battery?
    a. The thin filament is supported by two thicker conduction wires that are insulated from each other.
    b. Each end of the filament is connected to a different end of the battery.
    c. The filament is part of a complete circuit.
    * d. All of the above are true.

12. Electric current is measured in
    a. coulombs.
    * b. amperes.
    c. ohms.
    d. watts.

13. The electric current is a measure of the
    a. force exerted on the charges by the battery.
    * b. amount of charge flowing through a wire each second.
    c. energy supplied by the battery each second.
    d. forces resisting the flow of charge through the circuit.

14. The voltage in a circuit is a measure of the
    * a. force exerted on the charges by the battery.
    b. amount of charge flowing through a wire each second.
    c. energy supplied by the battery each second.
    d. resistance to the flow of charge through the circuit.

15. Which charges are free to move in metals?
    * a. negative
    b. positive
    c. both
    d. neither

16. Which charges are free to move in liquids?
    a. negative
    b. positive
    * c. both

17. Conventional current is the flow of
    a. negative charge.
    * b. positive charge.
    c. both types of charge.
    d. neither type of charge.

18. The current though a resistor is 4 A. The number of electrons flowing through the resistor in one minute is
    a. 1.60×10^{-19}.
    b. 240.
    c. 6.25×10^{18}.
    * d. 1.50×10^{21}.

19. The number of ohms in a circuit is a measure of the
    a. force exerted on the charges by the battery.
    b. amount of charge flowing through a wire each second.
    c. energy supplied by the battery each second.
    * d. resistance to the flow of charge through the circuit.

20. Which of the following does NOT affect the resistance of a wire?
    a. diameter
    b. type of metal
    c. length
    * d. voltage

21. Which of the following does NOT affect the resistance of a wire?
    * a. current
    b. type of metal
    c. length
    d. temperature

22. Increasing which of the following decreases the resistance of a wire?
    a. length
    b. temperature
    * c. diameter
    d. current

23. Which of the following combination of units is equivalent to an ohm?
    a. volt-ampere
    b. ampere/volt
    * c. volt/ampere
    d. volt^2/ampere

24. When resistors are connected in parallel, the resistances
    a. add.
    * b. add as reciprocals.
    c. stay the same.
    d. are averaged.

25. When resistors are connected in series, the resistances
    * a. add.
    b. add as reciprocals.
    c. stay the same.
    d. are averaged.

26. Which of the following is true of a superconducting material as it is steadily cooled to very low temperatures? Its resistance
    a. quickly drops to a very low value and remains there.
    b. drops steadily to a very low value.
    * c. drops to zero and remains there.

27. If the length of a wire is doubled, the resistance
    a. is reduced to one-fourth its original value.
    b. is cut in half.
    * c. doubles.
    d. quadruples.

28. If the radius of a wire is doubled, the resistance
    * a. is reduced to one-fourth its original value.
      b. is cut in half.
      c. doubles.
      d. quadruples.

29. What is the resistance of a 4-Ω resistor and a 12-Ω resistor connected in series?
      a. 3 Ω
      b. 4 Ω
      c. 8 Ω
      d. 12 Ω
    * e. 16 Ω

30. What is the resistance of a 4-Ω resistor and a 12-Ω resistor connected in parallel?
    * a. 3 Ω
      b. 4 Ω
      c. 8 Ω
      d. 12 Ω
      e. 16 Ω

31. What is the resistance of three 6-Ω resistors connected in series?
      a. 2 Ω
      b. 3 Ω
      c. 9 Ω
    * d. 18 Ω

32. What is the resistance of three 6-Ω resistors connected in parallel?
    * a. 2 Ω
      b. 3 Ω
      c. 9 Ω
      d. 18 Ω

33. A 100-Ω resistor is connected in parallel with a 50-Ω resistor. Which of the following best describes the equivalent resistance of this parallel network?
      a. greater than 150 Ω
      b. equal to 150 Ω
      c. equal to 75 Ω
    * d. less than 50 Ω

34. Three identical resistors, each 45 Ω, are connected in parallel across a 60-V battery. The current through the battery is
      a. 0.25 A.
      b. 0.75 A.
      c. 1.33 A.
  *   d. 4 A.

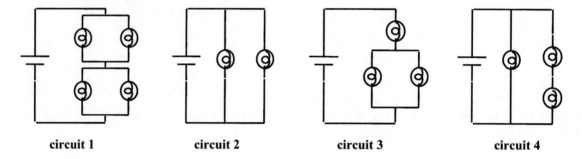

      circuit 1   circuit 2   circuit 3   circuit 4

35. Which of the circuits above has the least equivalent resistance?
      a. circuit 1
      b. circuit 2
      c. circuit 3
  *   d. circuit 4

36. Which of the circuits above has the most equivalent resistance?
      a. circuit 1
      b. circuit 2
  *   c. circuit 3
      d. circuit 4

37. If a 200-Ω resistor and a 400-Ω resistor are each connected to 12-V batteries, which resistor will draw the most current?
  *   a. 200-Ω
      b. 400-Ω
      c. They draw the same current.

38. A red resistor draws more current than a blue resistor when they are connected to the same battery. Which resistor has the lower resistance?
  *   a. red
      b. blue
      c. They have the same resistance.

Physics: A World View, Sixth Edition by Larry Kirkpatrick and Gregory Francis

39. What is the resistance of a light bulb that draws 2 A when it is plugged into a 120-V outlet?
  * a. 60 Ω
  b. 120 Ω
  c. 240 Ω
  d. 480 Ω

40. If a radio draws a current of 4 A when connected to 120 V, what is the resistance of its circuit?
  a. 7.5 Ω
  * b. 30 Ω
  c. 120 Ω
  d. 480 Ω

41. What voltage is needed to produce a current of 0.5 A in a 50-Ω resistor?
  a. 12 V
  * b. 25 V
  c. 500 V
  d. 100 V

42. Two identical resistors are connected in series across an ideal battery and draw 8 A through the battery. If the two resistors were instead connected in parallel across the battery, the current drawn through the battery would be
  a. 2 A.
  b. 4 A.
  c. 16 A.
  * d. 32 A.

43. A flashlight has four D-cells (1.5 V) placed end to end. What current will a 30-Ω bulb draw?
  a. 0.05 A.
  * b. 0.2 A.
  c. 5 A.
  d. 20 A.

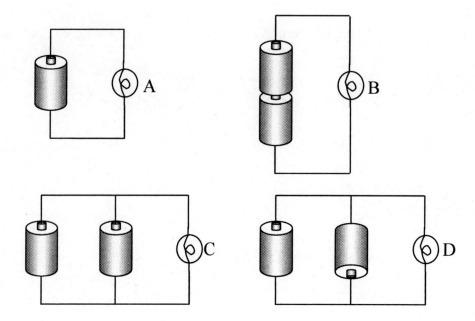

44. A single bulb is connected to each of the arrangements of identical batteries shown above. Which bulb will be the brightest?
     a. A
  * b. B
     c. C
     d. D

45. A single bulb is connected to each of the arrangements of identical batteries shown above. Which bulb will not light?
     a. A
     b. B
     c. C
  * d. D

46. A single bulb is connected to each of the arrangements of identical batteries shown above. Which bulb(s) will light, but be the dimmest?
     a. A
     b. B
  * c. A and C
     d. C

Physics: A World View, Sixth Edition by Larry Kirkpatrick and Gregory Francis

47. Which of the following options makes the statement that best describes a short circuit? _____ of the electricity flows through the short circuit.
     a. All
   * b. Almost all
     c. Half
     d. Almost none

48. Three identical bulbs are connected in series to a battery. Which bulb is the brightest?
     a. the one closest to the positive end of the battery
     b. the one closest to the negative end of the battery
     c. the middle one
   * d. The bulbs are equally bright.

49. To get the brightest bulbs with two batteries and two bulbs you would connect the batteries in _____ and the bulbs in _____.
     a. series ... series
     b. parallel ... series
   * c. series ... parallel
     d. parallel ... parallel

50. To get the dimmest bulbs with two batteries and two bulbs you would connect the batteries in _____ and the bulbs in _____.
     a. series ... series
   * b. parallel ... series
     c. series ... parallel
     d. parallel ... parallel

51. To get the longest lifetime from two batteries and two bulbs you would connect the batteries in _____ and the bulbs in _____.
     a. series ... series
   * b. parallel ... series
     c. series ... parallel
     d. parallel ... parallel

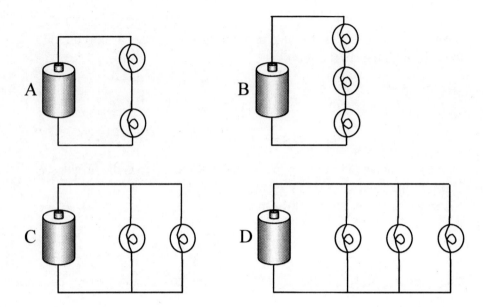

52. In which arrangement of bulbs shown above are the bulbs the brightest?
     a. A
     b. B and C
     c. B and D
  * d. C and D

53. In which arrangement of bulbs shown above are the bulbs the dimmest?
     a. A
  * b. B
     c. C
     d. D

54. In which arrangement of bulbs shown above will the battery last the longest?
     a. A
  * b. B
     c. C
     d. D

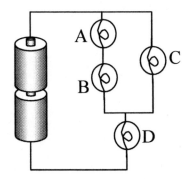

55. Which bulb is (or bulbs are) the brightest in the circuit shown above?
    a. A
    b. A and B
    c. C
    * d. D

56. What happens to the brightness of bulb A if D is removed in the circuit shown above?
    a. It becomes dimmer.
    b. It becomes brighter.
    c. It remains the same.
    * d. It goes out.

57. What happens to the brightness of bulb A if B burns out in the circuit shown above?
    a. It becomes dimmer.
    b. It becomes brighter.
    c. It remains the same.
    * d. It goes out.

58. What happens to the brightness of bulb B if a wire is connected across the two terminals of socket C in the circuit shown above?
    a. It becomes dimmer.
    b. It becomes brighter.
    c. It remains the same.
    * d. It goes out.

59. What happens to the brightness of bulb D if a wire is connected across the two terminals of socket C in the circuit shown above?
    a. It becomes dimmer.
    * b. It becomes brighter.
    c. It remains the same.
    d. It goes out.

60. What happens to the brightness of bulb D if C is removed from its socket (and the empty socket remains) in the circuit shown above?
    * a. It becomes dimmer.
      b. It becomes brighter.
      c. It remains the same.
      d. It goes out.

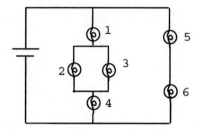

61. Six identical bulbs are connected to a battery as shown above. The proper ranking for the brightness of the bulbs is
      a. 1 = 4 > 5 = 6 > 2 = 3.
    * b. 5 = 6 > 1 = 4 > 2 = 3.
      c. 1 = 4 > 2 = 3 > 5 = 6.
      d. 1 = 4 > 5 = 6 = 2 = 3.

62. In a household circuit the lights, outlets, and appliances are wired in _____ with each other and in _____ with a fuse or circuit breaker.
      a. series ... series
    * b. parallel ... series
      c. series ... parallel
      d. parallel ... parallel

63. Three identical bulbs are connected in parallel across a 12-V battery. If one of the bulbs were removed from its socket then the potential difference measured across the empty socket would be
      a. zero.
      b. 4 V.
    * c. 12 V.
      d. infinite.

64. Which of the following is a unit for measuring electric power?
      a. joule
      b. coulomb
      c. ampere
    * d. watt

65. Which of the following is a unit for measuring electric energy?
    a. ampere
    b. volt
    c. watt
*   d. kilowatt-hour

66. Which is NOT a formula for calculating power?
    a. $P = \Delta E/\Delta t$
*   b. $P = IR$
    c. $P = IV$
    d. $P = I^2 R$

67. A flashlight has two D-cells (1.5 V) placed end to end. What current rating should the bulb have to deliver 6 W?
    a. 1.5 A
*   b. 2 A
    c. 3 A
    d. 3 V

68. A flashlight has four D-cells (1.5 V) placed end to end. What power rating should a 12-Ω bulb have?
    a. 0.33 W
*   b. 3 W
    c. 8 W
    d. 18 W

69. What is the power of a light bulb that draws 2 A when it is plugged into a 120-V outlet?
    a. 60 W
    b. 120 W
*   c. 240 W
    d. 480 W

70. What is the power of a light bulb that draws 0.5 A when it is plugged into a 120-V outlet?
*   a. 60 W
    b. 120 W
    c. 240 W
    d. 480 W

71. How much current does a 240-W light bulb draw when it is plugged into a 120-V outlet?
    - a. 1 A
    * b. 2 A
    - c. 4 A
    - d. 8 A

72. If electricity costs 10 cents/kWh, how much does it cost to run a 2-kW hair drier for one hour?
    - a. 1 cent
    - b. 2 cents
    - c. 10 cents
    * d. 20 cents

73. Which costs the least to operate for 24 hours?
    * a. a 5-W safety light
    - b. a 100-W light bulb
    - c. a 750-W curling iron
    - d. a 1500-W heater

74. Which costs the most to operate?
    * a. a 1500-W heater for 1 h
    - b. a 100-W light bulb for 8 h
    - c. a 40-W refrigerator for 20 h
    - d. a 250-W outdoor light for 2 h

Chapter 22: Electromagnetism

1. Magnetism is most closely related to
   a. sound.
   b. light.
   c. water waves.
   * d. electricity.

2. Which element does NOT occur naturally in the magnetized state?
   a. iron
   * b. zinc
   c. nickel
   d. cobalt

3. Which of the following does NOT apply to both electric and magnetic phenomena?
   a. Each has two types of "charge."
   b. Each can attract neutral objects.
   c. Each gets weaker with increasing distance.
   * d. "Charges" of each can be isolated.

4. If a bar magnet is broken into two pieces, each piece
   a. has only one magnetic pole.
   b. has an electric charge.
   * c. has two magnetic poles.
   d. is no longer magnetic.

5. Which statement about magnetic poles is NOT correct?
   a. Like magnetic poles repel; unlike magnetic poles attract.
   b. The strengths of the two poles of a magnet are the same.
   c. Both types of magnetic pole will attract unmagnetized iron.
   * d. Magnetic monopoles behave like isolated electric charges.

6. It is not possible for two magnetic field lines to cross because _____ at that point.
   a. two magnets cannot contribute to the field
   b. the field would have to be infinite
   * c. the field cannot have two different directions
   d. the magnetic potential energy would have to be infinite

7. The direction of the magnetic field at each point in space is defined to be the direction of the _____ a compass at that point.
   a. torque on
   * b. north end of
   c. south end of
   d. perpendicular to the direction of

Chapter 22   Electromagnetism

8. If you are given three iron rods, how could you use them to find the one that is not magnetized? The unmagnetized rod will
      a. be attracted to both magnetized rods.
      b. be attracted to one magnetized rod and repel the other one.
      c. attract one magnetized rod and repel the other.
  *   d. not be repelled by either of the magnetized rods.

9. You have three iron bars, each of which may or may not be a permanent magnet. Each rod is painted green on one end and yellow on the other. Which of the following results indicates that bar A must be a permanent magnet?
      a. The green end of bar A attracts the green end of bar B.
  *   b. The yellow end of bar A repels the green end of bar B.
      c. The green end of bar B repels the green end of bar C.

10. You have three iron bars, each of which may or may not be a permanent magnet. Each rod is painted green on one end and yellow on the other. You perform three experiments and find: The green end of bar A attracts the green end of bar B; the yellow end of bar A repels the green end of bar B; and the yellow end of bar B repels the yellow end of bar C. What is the interaction between the green end of bar A and the green end of bar C?
  *   a. attraction
      b. repulsion
      c. There is no interaction.

11. In the space surrounding a wire carrying a steady current, there exists
      a. a magnetic field pointing radially outward.
  *   b. a magnetic field circling the wire.
      c. an oscillating magnetic field.
      d. no magnetic field.

12. If the steady current in a wire is coming directly toward you, the magnetic field lines
      a. point radially outward.
      b. point radially inward.
      c. circle the wire in the clockwise direction.
  *   d. circle the wire in the counterclockwise direction.

13. If a circular ring in the plane of this page carries a steady current in the clockwise direction, the magnetic field at the center of the ring
  *   a. points into the page.
      b. points out of the page.
      c. is circular in the clockwise direction.
      d. is circular in the counterclockwise direction.

14. If you are looking in the end of a solenoid and observe the current going around the solenoid in the counter-clockwise direction, the magnetic field
  a. points into the solenoid.
  * b. points out of the solenoid.
  c. is circular in the clockwise direction.
  d. is circular in the counterclockwise direction.

15. Lodestone (magnetite) is an igneous rock, one that forms from molten material. How do you suppose it became magnetized?
  * a. It solidified in Earth's magnetic field.
  b. It was hit by other rocks that were magnetized.
  c. The heating process produced spontaneous magnetization.
  d. Electric currents in the molten material produced the magnetization.

16. Which is NOT a property of electromagnets?
  a. They can be turned on and off.
  b. The strength of their magnetic fields can be varied by varying the current.
  * c. The strength of their magnetic fields can be varied by varying their net charge.
  d. They can produce large magnetic fields.

17. Two parallel wires carrying currents in opposite directions will
  a. attract each other.
  * b. repel each other.
  c. cause one wire to move upward and the other downward.
  d. oscillate in standing waves.

18. If two parallel wires separated by one meter carry equal currents and the force on each meter of each wire is 2×10^{-7} newton, the current in each wire is one
  a. watt.
  b. coulomb.
  * c. ampere.
  d. volt.

19. A wire is lying between the jaws of a horseshoe-shaped magnet so that it is aligned perpendicular to the magnetic field. What is the direction of the force on the wire when it is connected to a battery?
  * a. perpendicular to both the magnetic field and the wire
  b. perpendicular to the magnetic field only
  c. perpendicular to the wire only
  d. There is no force exerted on the wire.

Chapter 22   Electromagnetism

20. The strength of the magnetic field at a distance of one meter from a long straight wire carrying a current of one ampere is 2×10^{-7}
      a. ampere-meter.
      b. gauss.
  * c. tesla.
      d. volt per meter.

21. The geographic point in Northern Canada that we call the North Magnetic Pole is really a
      a. positive charge.
      b. negative charge.
      c. magnetic north pole.
  * d. magnetic south pole.

22. The most likely source of Earth's magnetic field is
      a. magnetized iron and nickel at Earth's center.
  * b. electric currents circulating in Earth's interior.
      c. a deposit of magnetized iron located just north of Hudson Bay.
      d. charged particles circulating in Earth's upper atmosphere.

23. Geologic evidence indicates that Earth's magnetic field has
      a. steadily gotten weaker.
      b. had a nearly constant value.
      c. steadily gotten stronger.
  * d. reversed many times.

24. A magnet produces a magnetic field that points vertically upward. What is the direction of the force on a proton that enters this region with a horizontal velocity toward the north?
      a. along the north-south direction
  * b. along the east-west direction
      c. along the up-down direction

25. An electron has a velocity of 5×10^6 m/s perpendicular to a magnetic field of 1.5 T. What force does the electron experience?
  * a. 1.2×10^{-12} N
      b. 3.0×10^{-7} N
      c. 3.3×10^6 N
      d. 7.5×10^6 N

26. A proton with a speed v perpendicular to a magnetic field B experiences a force F. If the speed of the proton is doubled, the new force is
        a. 0.5 F.
        b. still F.
    *   c. 2 F.
        d. 4 F.

27. An electron with a speed v perpendicular to a magnetic field B experiences a force F. If the strength of the magnetic field is cut in half, the new force is
    *   a. 0.5 F.
        b. still F.
        c. 2 F.
        d. 4 F.

28. An electron with a speed v perpendicular to a magnetic field B experiences a force F that causes it to move along a circular path of radius r. If the strength of the magnetic field is doubled, the new radius is
    *   a. 0.5 r
        b. still r
        c. 2 r
        d. 4 r

    Difficulty: 2

29. An electron with a speed v perpendicular to a magnetic field B experiences a force F that causes it to move along a circular path of radius r. If the speed of the electron is doubled, the new radius is
        a. 0.5 r.
        b. still r.
    *   c. 2 r.
        d. 4 r.

30. In which of the following situations is an electric current NOT produced?
        a. A magnet moves relative to a stationary wire.
        b. A wire moves relative to a stationary magnet.
    *   c. A wire has been wrapped around a stationary magnet.
        d. The magnetic field through a loop is decreasing.

Chapter 22   Electromagnetism

31. In which of the following situations is an electric current NOT produced?
      a. A loop of wire rotates in a steady magnetic field.
  * b. A loop of wire is stationary in a steady magnetic field.
      c. The magnetic field through a loop is increasing.
      d. The magnetic field through a loop is decreasing.

32. A magnetic field does NOT exert a force on a
      a. paper clip.
      b. magnet.
  * c. stationary charge.
      d. moving charge.

33. Quickly inserting the north end of a bar magnet into a coil of wire causes the needle of a meter to deflect to the right. Which of the following will also produce a deflection to the right?
      a. inserting a south pole into the coil
      b. removing a north pole from the coil
  * c. removing a south pole from the coil
      d. None of these will produce a deflection to the right.

34. A conducting loop is lying flat on the ground. The north pole of a bar magnet is brought down toward the loop. As the magnet approaches the loop, there will be an induced
      a. current clockwise around the loop.
  * b. current counter-clockwise around the loop.
      c. voltage but no induced current around the loop.

35. A conducting loop is lying flat on the ground. The south pole of a bar magnet is brought down toward the loop. As the magnet approaches the loop, there will be an induced
  * a. current clockwise around the loop.
      b. current counter-clockwise around the loop.
      c. voltage but no induced current around the loop.

36. A conducting loop is lying flat on the ground. A bar magnet is held directly above the loop with the south pole of the magnet closest to the loop. The magnet is then raised quickly away from the loop. As the magnet leaves the loop, there will be an induced
      a. current clockwise around the loop.
  * b. current counter-clockwise around the loop.
      c. voltage but no induced current around the loop.

37. A conducting loop is lying flat on the ground. A bar magnet is held directly above the loop with the north pole of the magnet closest to the loop. The magnet is then raised quickly away from the loop. As the magnet leaves the loop, there will be an induced
    * a. current clockwise around the loop.
    b. current counter-clockwise around the loop.
    c. voltage but no induced current around the loop.

38. A copper ring is oriented perpendicular to a uniform magnetic field. If the ring is suddenly moved in the direction opposite the field lines, the magnitude of the net magnetic field in the center of the loop (the uniform field plus the induced field) will be _____ the magnitude of the uniform field?
    a. greater than
    * b. equal to
    c. less than

39. A copper ring is oriented perpendicular to a uniform magnetic field. The ring is quickly stretched such that its radius doubles over a short time. As the ring is being stretched, the magnitude of the net magnetic field in the center of the loop (the uniform field plus the induced field) will be ____ the magnitude of the uniform field?
    a. greater than
    b. equal to
    * c. less than

40. A transformer CANNOT be used to
    a. increase the voltage of ac electricity.
    b. decrease the voltage of ac electricity.
    c. increase the current of ac electricity.
    * d. increase the current of dc electricity.

41. A transformer can be used to increase the
    * a. voltage of ac electricity.
    b. power of ac electricity.
    c. voltage of dc electricity.
    d. power of dc electricity.

42. A transformer with 100 turns in the primary coil and 400 turns in the secondary coil will change 120-V ac electricity into _____ ac electricity.
    a. 30-V
    b. 60-V
    c. 240-V
    * d. 480-V

Chapter 22   Electromagnetism

43. A transformer is used to convert 120-V household electricity to 12 V for use in a CD player. If the primary coil connected to the outlet has 400 loops, how many loops does the secondary coil have?
    a. 20
    * b. 40
    c. 2000
    d. 4000

44. The voltage in the lines that carry electric power to homes is typically 2000 V. What is the required ratio of the loops in the primary and secondary coils of the transformer to drop the voltage to 120 V?
    a. 10:1
    * b. 16.7:1
    c. 1:10
    d. 1:16.7

45. A transformer is used to step down the voltage from 120 V to 6 V for use with an electric razor. If the razor draws a current of 0.5 A, what current is drawn from the 120-V line?
    * a. 0.025 A
    b. 0.5 A
    c. 3 A
    d. 10 A

46. If a loop of wire is rotated in a steady magnetic field, the voltage produced in the loop will be a maximum when the plane of the loop is
    a. perpendicular to the field.
    * b. parallel to the field.
    c. at 45° to the field.

47. If a loop of wire is rotated in a steady magnetic field, the voltage produced in the loop will be zero when the plane of the loop is
    * a. perpendicular to the field.
    b. parallel to the field.
    c. at 45° to the field.
    d. The voltage will never be zero.

48. What type of electricity is produced by rotating a coil at a steady rate in a steady magnetic field?
    * a. ac electricity
    b. dc electricity
    c. pulsed dc electricity
    d. steady current

49. Which of the following statements is NOT correct as written?
       a. A changing electric field can produce a magnetic field.
   *   b. A constant magnetic field can produce a current.
       c. A changing magnetic field can produce a current.
       d. A changing magnetic field can produce an electric field.

50. Which of the following statements is NOT correct as written?
       a. A changing electric field can produce a changing magnetic field.
   *   b. A steady magnetic field produces a steady current.
       c. A changing magnetic field can produce a changing current.
       d. A changing magnetic field can produce a steady electric field.

51. Each of the following statements is correct as written. Which one is NOT true if you interchange the underlined words in each sentence?
       a. A changing <u>magnetic</u> field can produce a changing <u>electric</u> field.
   *   b. A steady <u>current</u> produces a steady <u>magnetic field</u>.
       c. A changing <u>current</u> can produce a changing <u>magnetic field</u>.
       d. A changing <u>electric</u> field can produce a steady <u>magnetic</u> field.

52. Which of the following statements is NOT true for electromagnetic waves?
       a. They travel in a vacuum at the speed of light.
       b. They are generated by accelerating electric charges.
       c. They are oscillating electric and magnetic fields traveling through space.
   *   d. Only a restricted range of frequencies can exist.

53. Which of the following are NOT electromagnetic waves?
       a. radio
       b. TV
       c. infrared light
   *   d. sound

54. Which of the following electromagnetic waves has the highest frequency?
   *   a. X rays
       b. radio
       c. microwaves
       d. ultraviolet light

55. What is the frequency of a microwave with a wavelength of 1 cm?
       a. 3×10^7 Hz
       b. 3×10^8 Hz
       c. 3×10^9 Hz
   *   d. 3×10^{10} Hz

Chapter 22   Electromagnetism

56. What is the frequency of an X ray with a wavelength of 1 nm?
    a. 3 Hz
    b. 30 Hz
 * c. 3×10^{17} Hz
    d. 3×10^{18} Hz

57. What is the wavelength of an X ray with a frequency of 3×10^{18} Hz?
    a. 1 mm
    b. 0.1 mm
    c. 0.1 μm
 * d. 0.1 nm

58. The signal broadcast by an AM radio station
    a. is a direct, electromagnetic replication of the sound.
 * b. has the amplitude of the carrier signal modulated by the sound.
    c. has the frequency of the carrier signal modulated by the sound.
    d. is the same as that broadcast by a FM station.

59. The signal broadcast by a FM radio station
    a. is a direct, electromagnetic replication of the sound.
    b. has the amplitude of the carrier signal modulated by the sound.
 * c. has the frequency of the carrier signal modulated by the sound.
    d. is the same as that broadcast by an AM station.

60. What is the wavelength of the carrier wave for an AM radio station located at 1000 on the dial?
    a. 3 cm
    b. 3 m
    c. 30 m
 * d. 300 m

Chapter 23: The Early Atom

1. The periodic table arranges the elements according to
   - a. the order in which they were discovered.
   - * b. their chemical properties.
   - c. their relative abundances.
   - d. alphabetical order.

2. What allowed Mendeleev to predict the existence of new elements?
   - a. a larger than usual gap in the atomic masses
   - b. a change in the pattern for the natural abundances of the elements
   - * c. gaps when the elements were arranged by their chemical similarities
   - d. a change in the ratios given by the law of definite proportions

3. A gas can be identified by means of its spectral lines because each element
   - a. can be recognized when magnified greatly.
   - b. occupies a unique position in the periodic table.
   - * c. emits characteristic wavelengths when electrically excited.
   - d. has a different atomic mass.

4. A beam of white light enters from the east side of a glass container filled with a cool gas. The beam leaves the container on the west side, and is then passed through a prism to create a spectrum. The spectrum observed is known as _____ spectrum.
   - a. a continuous (rainbow)
   - b. an emission (bright line)
   - * c. an absorption (dark line)
   - d. a gaseous

5. A beam of white light enters from the east side of a glass container filled with a cool gas. The beam leaves the container on the west side. Light from the container that leaves on the north side is passed through a prism to produce a spectrum. The spectrum observed is known as _____ spectrum.
   - a. a continuous (rainbow)
   - * b. an emission (bright line)
   - c. an absorption (dark line)
   - d. a gaseous

6. The spectral lines for elements are called "atomic fingerprints" because they
   - a. resemble human fingerprints.
   - * b. are unique for each element.
   - c. leave impressions on photographic film.
   - d. are colorful.

7. The number of lines in the emission spectrum for an element is _____ that in the absorption spectrum.
    a. the same as
 * b. greater than
    c. less than
    d. The two spectra cannot be compared.

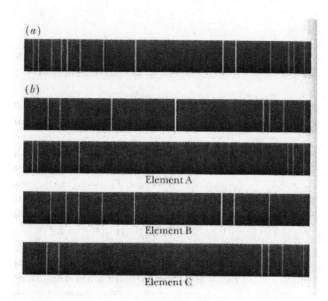

Note: This is line art from page 495 of the text.

8. Which of the elements A, B, and C are present according to the spectrum labeled *a* in the diagram shown above?
    a. only A
    b. only B
    c. only C
 * d. A and B

9. Which of the elements A, B, and C are present according to the spectrum labeled *b* in the diagram shown above?
    a. only A
    b. only B
 * c. only C
    d. B and C

10. Cathode rays are shown NOT to be electromagnetic radiation by the observation that they
      a. travel in straight lines.
      b. cast shadows.
      c. exist only in a vacuum.
  *   d. are deflected by electric and/or magnetic fields.

11. Which is a correct observation of what happens in a cathode ray tube?
      a. The end of the tube near the cathode glows.
  *   b. A metal plate casts a shadow.
      c. The particles are emitted from the anode.
      d. The particles are not deflected by an electric field.

12. In his studies of cathode rays, J.J. Thomson
      a. showed that their charge-to-mass ratio depended on the cathode material.
      b. used magnetic fields to show that cathode rays have positive charges.
      c. determined that their charge-to-mass ratio was smaller than for hydrogen ions.
  *   d. used electric fields to show that cathode rays have negative charges.

13. Cathode rays are
      a. negatively charged.
      b. electrons.
      c. pieces of atoms.
  *   d. All of the above.

14. An electron from a hydrogen atom _____ an electron from an oxygen atom.
  *   a. is identical to
      b. has a greater mass than
      c. is larger than
      d. has a greater charge than

15. The charge-to-mass ratio for cathode rays is
      a. 1800 times smaller than that for hydrogen ions.
      b. 1800 times smaller than that for electrons.
  *   c. 1800 times larger than that for hydrogen ions.
      d. equal to that for hydrogen ions.

16. The charge-to-mass ratio for cathode rays is _____ that for electrons.
      a. smaller than
  *   b. equal to
      c. larger than

Chapter 23   The Early Atom

17. The charge-to-mass ratio for cathode rays is _____ that for hydrogen ions.
    a. smaller than
    b. equal to
    * c. larger than

18. In his famous oil-drop experiment Robert Millikan discovered that
    a. not all electrons are the same.
    b. cathode rays are electromagnetic waves.
    * c. electrical charge comes in identical chunks, or quanta.
    d. cathode rays are electrons.

19. The Thomson model of the atom explained why
    a. atoms look like miniature solar systems.
    b. positive charges are emitted from cathodes.
    * c. all materials emit electrons that are identical.
    d. the chemical elements exhibit a periodicity.

20. In the Rutherford model of the atom nearly all the volume occupied by matter consists of
    a. electrons.
    b. protons.
    c. neutrons.
    * d. a vacuum.

21. Which one of the following lists has the most massive object listed first and the least massive last?
    a. nucleus, electron, atom
    b. atom, electron, nucleus
    * c. atom, nucleus, electron
    d. nucleus, atom, electron

22. What force is responsible for holding the electrons in their orbits about nuclei?
    a. gravitational
    * b. electric
    c. magnetic
    d. No force is required.

23. Rutherford's model of an atom as a very tiny, massive nucleus with the electrons orbiting at great distances could NOT explain why
    a. most alpha particles pass right through a thin gold foil.
    b. some alpha particles are deflected at large angles, even backward.
    * c. atoms emit a set of discrete wavelengths.

24. The Rutherford model of the atom was successful in explaining
    * a. the scattering of alpha particles from the gold foil.
    b. the unique spectral lines for each element.
    c. the stability of atoms.
    d. the periodicity of the chemical elements.

25. Rutherford's model predicted that atoms should be unstable (the electrons should spiral into the nucleus) over very short time periods. What caused this instability in Rutherford's model?
    a. The positive charge in the nucleus was too far from the electrons to hold them in orbit.
    b. The attractive force between the positive nucleus and the electrons would pull them together.
    * c. Accelerating charge must radiate energy.
    d. Nature abhors a vacuum.

26. The curves for the intensities of the colors emitted by hot objects
    a. give atomic fingerprints of the materials.
    * b. change with temperature but not appreciably with material.
    c. have far greater intensities in the ultraviolet than the infrared for all temperatures.
    d. have far greater intensities in the ultraviolet than the infrared for all materials.

27. As the temperature of a radiating object increases, the wavelength with the maximum intensity
    a. increases.
    b. stays the same.
    * c. decreases.

28. As the temperature of a radiating object increases, the frequency with the maximum intensity
    * a. increases.
    b. stays the same.
    c. decreases.

29. As the temperature of a radiating object increases, the intensity of the radiation
    * a. increases.
    b. stays the same.
    c. decreases.

30. Planck was able to obtain the correct curve for the spectrum of light emitted by a hot object by assuming that the
    a. spectrum was continuous.
    * b. vibrational energies of the atomic resonators were quantized.
    c. atoms had only certain allowed orbits.
    d. light was mainly ultraviolet.

31. You measure the brightness of two different hot objects; first with a blue filter and then with a red filter. You find that object A has a brightness of 25 in the blue and 20 in the red. Object B has a brightness of 12 in the blue and 3 in the red. The brightness units are arbitrary but the same for all measurements. Object A is _____ object B.
    *   a. cooler than
        b. the same temperature as
        c. hotter than

32. What color star is the hottest?
        a. red
        b. orange
        c. white
    *   d. blue

33. The energy of one of Planck's quanta can be calculated using
        a. $E = h\lambda$.
        b. $E = h/\lambda$.
    *   c. $E = hf$.
        d. $E = h/f$.

34. A photon of yellow light has a wavelength of 6×10^{-7} m. What is its energy? Planck's constant is 6.63×10^{-34} J·s.
        a. 3.98×10^{-40} J
        b. 1.11×10^{-27} J
    *   c. 3.32×10^{-19} J
        d. 9.05×10^{26} J

35. A microwave photon has an energy of 2×10^{-23} J. What is its wavelength? Planck's constant is 6.63×10^{-34} J·s.
        a. 1.33×10^{-56} m
        b. 3.32×10^{-11} m
    *   c. 9.95×10^{-3} m
        d. 3.02×10^{10} m

36. Photoelectrons are emitted from a metallic surface only when the incident light has more than a certain minimum
        a. speed.
        b. wavelength.
    *   c. frequency.
        d. intensity.

37. When light is incident on a metallic surface, the emitted electrons
    a. are called photons.
    b. have arbitrarily high energies.
    c. have a maximum energy that depends on the intensity of the light.
    * d. have a maximum energy that depends on the frequency of the light.

38. Einstein was able to account for the experimental observations of the photoelectric effect by assuming that
    a. the metal contained atomic resonators.
    b. light is a wave phenomenon.
    * c. light consists of photons.
    d. electrons boil off when they get hot enough.

39. What color light will produce photoelectrons with the highest average kinetic energy?
    a. red
    * b. blue
    c. yellow
    d. green

40. A clean surface of potassium metal will emit electrons when exposed to blue light. If the intensity of the blue light is increased, the _____ of the ejected electrons will also increase.
    a. maximum kinetic energy
    * b. number
    c. average speed
    d. mass

41. A clean surface of metal will emit electrons when exposed to light. If the color of the light is changed from red to blue without changing the intensity, the _____ of the ejected electrons will also increase.
    * a. maximum kinetic energy
    b. number
    c. mass
    d. charge

42. You find that if you shine ultraviolet light on a negatively charged electroscope, the electroscope discharges. A positively charged electroscope can be discharged with
    a. red light.
    b. any color light.
    c. ultraviolet light.
    * d. You cannot discharge a positively charged electroscope by shining light on it.

43. Which of the following lists photons in order of increasing energy?
        a. X-ray, radio, infrared, visible, ultraviolet
        b. infrared, visible, ultraviolet, X-ray, radio
        c. radio, infrared, X-ray, visible, ultraviolet
     * d. radio, infrared, visible, ultraviolet, X-ray

44. Which of the following is NOT a feature of the Bohr model of the atom?
     * a. an electron probability cloud
        b. electrons in planetary-like orbits
        c. quantized energy levels
        d. accelerating electrons that do not radiate

45. Bohr gave the following reason for the electron in the hydrogen atom existing only in certain discrete energy levels.
        a. This agrees with Newtonian mechanics.
        b. This agrees with Maxwell's equations.
        c. Both of the above were cited.
     * d. He simply postulated it.

46. In the Bohr model, a hydrogen atom is in its ground state when its electron
        a. is at the center of the atom.
        b. has been ionized.
     * c. is in the innermost orbit.
        d. has absorbed a photon.

47. Which of the following is quantized in the Bohr model?
        a. radius
        b. angular momentum
        c. energy
     * d. All of these are quantized.

48. Which of the following is NOT a postulate of the Bohr model?
     * a. The electron forms a standing wave about the nucleus.
        b. The angular momentum of the electron is quantized.
        c. The electron has a constant energy in the allowed orbits.
        d. A photon is emitted or absorbed when the electron changes orbits.

49. An electron in the ground state of the Bohr atom has a radius of 0.053 nm. What is the radius of the first excited state?
    - a. 0.053 nm
    - b. 0.106 nm
    - c. 0.159 nm
    - * d. 0.212 nm

50. The volume of the hydrogen atom in the excited $n = 2$ state is ___ times bigger than the volume of the hydrogen atom in the ground state.
    - a. 2
    - b. 4
    - c. 16
    - * d. 64

51. What is the quantum number of the orbit in the hydrogen atom that has 36 times the radius of the smallest orbit?
    - * a. 6
    - b. 12
    - c. 18
    - d. 36

52. The angular momentum of the electron in the ground state of hydrogen is 1.06×10^{-34} J·s. What is the angular momentum of the electron in the $n = 4$ excited state of hydrogen?
    - a. 1.06×10^{-34} J·s
    - b. 2.12×10^{-34} J·s
    - * c. 4.24×10^{-34} J·s
    - d. 17×10^{-34} J·s

53. Two hydrogen atoms have electrons in the $n = 3$ energy level. One of the electrons jumps to the $n = 2$ level, while the other jumps to the $n = 1$ level. Which property is the same for the two photons that are emitted?
    - * a. velocity
    - b. frequency
    - c. energy
    - d. color

54. Two hydrogen atoms have electrons in the $n = 3$ energy level. One of the electrons jumps to the $n = 2$ level, while the other jumps to the $n = 1$ level. Which property is larger for the first photon?
      a. velocity
      b. frequency
  * c. wavelength
      d. energy

55. Two hydrogen atoms have electrons that jump from the $n = 3$ energy level to the $n = 1$ level. One jumps directly to the $n = 1$ level emitting one photon, while the other jumps to the $n = 2$ level first and then to the $n = 1$ level, emitting two photons. The total energy of the pair of photons is _____ that of the single photon.
      a. greater than
  * b. the same as
      c. less than
      d. The two energies cannot be compared.

56. The number of spectral lines in the absorption spectrum is _____ the number in the emission spectrum because _____
      a. equal to ... they come from the same kind of atom.
  * b. smaller than ... the electrons are almost all in the ground state.
      c. smaller than ... the energy of the photons must be constant.
      d. larger than ... the energy of the photons must be constant.

57. If electrons in hydrogen atoms are excited to the fourth Bohr orbit, how many different frequencies of light may be emitted?
      a. 1
      b. 3
  * c. 6
      d. 8

58. Certain families of elements in the periodic table have very similar properties because in each family
      a. the atoms have the same number of electrons.
      b. the masses of the atoms are simple multiples of each other.
  * c. the outermost electrons are in the same configuration.
      d. they evolved from the same basic elements.

59. Sulfur is element #16. How many electrons do you expect to find in each shell of a sulfur atom in its ground state?
    * a. 2, 8, 6
      b. 4, 4, 4, 4
      c. 2, 4, 6, 4
      d. 2, 7, 7

60. Roentgen discovered that X rays
      a. are high-energy electrons.
      b. can penetrate only a few centimeters of air.
      c. are deflected by electric and/or magnetic fields.
    * d. can produce fluorescence.

61. The spikes in the X-ray spectrum are due to
      a. electrons slowing down in the material.
      b. electrons knocked from the outer shells.
    * c. photons emitted by electrons dropping to fill inner shells.
      d. photons absorbed by electrons in the ground state.

62. Bohr's model predicts that the energies of an element's characteristic X rays
    * a. increase with atomic number.
      b. decrease with atomic number.
      c. increase with atomic mass.
      d. decrease with atomic mass.

Chapter 24: The Modern Atom

1. Which one of the following is considered to be a success of Bohr's theory of the atom?
    a. explaining the relative intensities of the spectral lines
    b. accounting for the splitting of some of the spectral lines
    * c. obtaining the numerical values for the spectral lines in hydrogen
    d. explaining why electrons in fixed orbits do not radiate

2. Which of the following is NOT considered to be a success of Bohr's theory of the atom?
    a. obtaining the numerical values for the spectral lines in hydrogen
    b. explaining why there are more lines in emission spectra than absorption spectra
    * c. explaining why electrons in fixed orbits do not radiate
    d. providing the general features of the periodic table

3. Which of the following technical terms CANNOT be used to describe both an electron and a photon?
    a. wavelength
    * b. mass
    c. energy
    d. momentum

4. Bohr could never really explain why an electron was limited to certain orbits. De Broglie explained this by showing that electrons

    * a. form standing-wave patterns about the nucleus.
    b. have elliptical orbits like the planets around the Sun.
    c. occupy a continuum of orbits but only radiate from some.
    d. obey Maxwell's equations.

5. You find that the lowest frequency at which you can set up a standing wave in a wire loop is 10 Hz. When you increase the driving frequency slightly above 10 Hz, the resonance goes away. What is the next frequency at which resonance will again appear?
    a. 15 Hz
    * b. 20 Hz
    c. 30 Hz
    d. 40 Hz

6. The wave behavior of bowling balls is not observed because
    a. their speeds are too small.
    b. their momenta are too small.
    * c. their wavelengths are too small.
    d. wave properties only occur on the atomic scale.

7. Your sports car does NOT diffract off the road when you drive through a tunnel because
    a. wave properties only occur for atomic-sized particles.
    b. the waves of the particles in the car interfere destructively.
    c. the waves of the particles in the car interfere constructively.
    *   d. the wavelength of the car is very small compared to the width of the tunnel.

8. As the speed of a particle increases, the de Broglie wavelength of the particle
    a. increases.
    *   b. decreases.
    c. stays the same.
    d. The wavelength is not defined.

9. If a baseball and a BB have the same speed, which has the shorter de Broglie wavelength?
    *   a. the baseball
    b. the BB
    c. They have the same wavelength.
    d. The wavelength is not defined.

10. In 1923, the French graduate student Louis de Broglie proposed that
    a. photons behave like particles.
    *   b. electrons behave like waves.
    c. the energy levels in atoms are quantized.
    d. the behavior of electrons must be explained by quantum mechanics.

11. Confirmation of de Broglie's idea that particles behave like waves came from an experiment in which
    a. atomic spectra were measured very precisely.
    *   b. electrons were scattered from nickel crystals.
    c. ultraviolet light illuminated a polished metal surface.
    d. light was observed to scatter from isolated electrons.

12. The de Broglie wavelength of a particle with a mass m and a velocity v is given by
    a. mv.
    b. λ/mv.
    c. mv/h.
    *   d. h/mv.

13. What is the de Broglie wavelength of a Volkswagen (mass = 1000 kg) traveling at 30 m/s (67 mph)? Planck's constant is 6.63 x 10^{-34} J·s.
      a. 1.47 x 10^{-39} m
  * b. 2.21 x 10^{-38} m
      c. 1.99 x 10^{-29} m
      d. 2.98 x 10^{-28} m

14. A bullet for a 30-06 rifle has a mass of 10 g and a muzzle velocity of 900 m/s. What is its de Broglie wavelength? Planck's constant is 6.63 x 10^{-34} J·s.
      a. 7.37 x 10^{-37} m
      b. 1.64 x 10^{-37} m
  * c. 7.37 x 10^{-35} m
      d. 5.97 x 10^{-30} m

15. When we perform the two-slit experiment with electrons, we find that the electrons behave
      a. just like bullets.
      b. just like water waves.
  * c. like bullets when they are detected but the pattern is wave-like.
      d. like waves when they are detected but the pattern is particle-like.

16. When we perform the two-slit experiment with photons, we find that the photons behave
      a. just like bullets.
      b. just like water waves.
  * c. like bullets when they are detected but the pattern is wave-like.
      d. like waves when they are detected but the pattern is particle-like.

17. If the two-slit experiment is performed with a beam of electrons so weak that only one electron passes through the apparatus at a time, what kind of pattern would you expect to obtain on the detecting screen?
      a. one hump
      b. two humps showing no interference
  * c. an interference pattern
      d. the sum of the patterns formed when each slit is open by itself.

18. If the two-slit experiment is performed with a beam of photons so weak that only one photon passes through the apparatus at a time, what kind of pattern would you expect to obtain on the detecting screen?
      a. one hump
      b. two humps showing no interference
  * c. an interference pattern
      d. the sum of the patterns formed when each slit is open by itself.

Physics: A World View, Sixth Edition by Larry Kirkpatrick and Gregory Francis

19. De Broglie waves can be regarded as _____ waves.
      a. pressure
  * b. probability
      c. electromagnetic
      d. gravitational

20. If an atomic particle is confined to a box and if it has the lowest possible energy, where would you most likely find the particle?
  * a. in the center
      b. near an edge
      c. any place
      d. midway between the center and an edge

21. If an atomic particle is confined to a box and if it is in the first excited energy state, where would you most likely find the particle?
      a. in the center
      b. near an edge
      c. any place
  * d. midway between the center and an edge

22. Which of the following properties of a particle in a box is NOT quantized?
      a. wavelength
      b. energy
  * c. position
      d. momentum

23. Which of the following is NOT a feature of the quantum-mechanical model of the atom?
      a. quantized energy levels
      b. discrete spectral lines
      c. an electron probability cloud
  * d. electrons in planetary-like orbits

24. The quantum-mechanical model of the atom does NOT have the problem of accelerating charges emitting electromagnetic radiation because the electrons
  * a. behave like waves and do not have well-defined orbits.
      b. do not have a charge.
      c. are stationary.
      d. have paths for which there is no acceleration.

25. Which of the following statements does NOT describe the quantum-mechanical model of the atom?

        a. The nucleus is surrounded by an electron cloud.

  *   b. The electrons orbit the nucleus in elliptical paths.

        c. The energy levels in the atom are quantized.

        d. Each electron is described by four quantum numbers.

26. Some minerals glow when they are illuminated with ultraviolet light. This occurs because

        a. the ultraviolet photons have lower energies than visible light.

        b. certain colors in the incident light are reflected.

        c. the ultraviolet light has a high intensity.

  *   d. the electrons cascade through several energy levels on the way to the ground state.

27. Some minerals continue to glow after the illuminating light is turned off. This occurs because

        a. the ultraviolet photons have lower energies than visible light.

  *   b. some of the electrons remain in the excited state for relatively long times.

        c. the ultraviolet light has a high intensity.

        d. the electrons jump directly to the ground state.

28. Which of the following types of electromagnetic radiation can cause fluorescence or phosphorescence?

        a. microwave

        b. infrared

        c. visible

  *   d. ultraviolet

29. The simple act of confining a particle to a three-dimensional box results in _____ quantum numbers.

        a. 1

        b. 2

  *   c. 3

        d. 4

30. The Pauli exclusion principle says that

        a. we cannot determine the position and momentum of a particle along a given direction with arbitrary accuracy.

        b. electrons behave like particles and waves.

        c. questions about the nature of electrons are excluded from discussions in science.

  *   d. no two electrons can have the same set of quantum numbers.

31. How many quantum numbers are required to specify the state of an atomic electron?
    a. 1
    b. 2
    c. 3
    * d. 4

32. Certain families of elements in the periodic table have very similar properties because in each family
    a. the atoms have the same number of electrons.
    b. the masses of the atoms are simple multiples of each other.
    * c. the outermost electrons are in the same configuration.
    d. they evolved from the same basic elements.

33. What is the range for the integral values of the component of the angular momentum along any defined direction?
    a. 0 to n, where n is the energy quantum number.
    b. 0 to $n-1$, where n is the energy quantum number.
    c. 0 to ℓ, where ℓ is the angular momentum quantum number.
    * d. $-\ell$ to $+\ell$, where ℓ is the angular momentum quantum number.

34. How many electrons occupy the lowest energy shell in nitrogen atoms?
    a. 1
    * b. 2
    c. 4
    d. 8

35. Given that chlorine is element #17, how many electrons occupy each of the first three energy levels in chlorine starting with the lowest?
    a. 7, 8, 2
    * b. 2, 8, 7
    c. 2, 6, 9
    d. 6, 6, 5

36. How many quantum states are there with $n = 3$?
    a. 6
    b. 8
    c. 12
    * d. 18

37. How many quantum states are there with $n = 4$?
    a. 16
    b. 18
    c. 24
    * d. 32

38. How many electrons can have the quantum numbers $n = 5$ and $\ell = 1$?
    a. 2
    b. 4
    c. 5
    * d. 6

39. How many electrons can have the quantum numbers $n = 5$ and $\ell = 4$?
    a. 2
    b. 9
    * c. 18
    d. 20

40. The Heisenberg uncertainty principle says that
    a. the human mind cannot cope with all the data needed to predict experimental results with absolute precision.
    b. no experimental property can be determined with absolute precision.
    * c. the position and momentum of a particle along a given direction cannot be determined with arbitrary accuracy.
    d. there are too many atoms in matter to be able to predict experimental results with arbitrary accuracy.

41. According to the Heisenberg uncertainty principle, it is impossible to precisely determine an object's position and its _____ at the same time.
    a. energy
    * b. momentum along the same direction
    c. momentum along the perpendicular direction
    d. angular momentum

42. According to the Heisenberg uncertainty principle, it is impossible to precisely determine an object's
    a. energy.
    b. life time.
    c. position.
    * d. energy and life time at the same time.

43. What quantum-mechanical variable is complementary to time?
    a. position
    b. momentum
 * c. energy
    d. angular momentum

44. The Heisenberg uncertainty principle is concerned with the fact that
    a. technology has not yet learned how to answer some of science's fundamental questions.
    b. one person can know an answer about which another person is uncertain.
    c. we cannot record enough data to accurately predict outcomes of experiments.
 * d. there is a limit to our ability to measure quantities due to the fundamental nature of the atomic world.

45. The complementarity principle was suggested by Bohr to
    a. get physicists to be kinder to each other when debating ideas.
    b. explain why there is a lower limit to our knowledge of the atom.
 * c. point out that a complete description of an electron requires wave and particle aspects.
    d. explain the periodicity of the chemical elements.

46. Scientist believe in the complementarity principle because
    a. it was stated by a famous physicist.
    b. western scientific thought is moving toward eastern religious thought.
    c. two theories are better than one.
 * d. both particle and wave descriptions are needed to understand the behaviors of electrons and photons.

47. If you could determine which slit the electron goes through, what effect would if have on the two-slit interference pattern?
    a. It would have no effect.
    b. It would reduce the overall intensity of the pattern.
 * c. It would destroy the interference pattern.
    d. It would become a superposition of two single-slit diffraction patterns.

48. "The quantum mechanics is very imposing. But an inner voice tells me that it is still not the final truth. The theory yields much, but it hardly brings us nearer to the secret of the Old One. In any case, I am convinced that He does not throw dice." What about quantum mechanics was troubling Albert Einstein?
    a. its mathematical sophistication
 * b. its probabilistic nature
    c. its great differences from Newtonian physics
    d. its wave-particle dualism

Chapter 24   The Modern Atom

49. Which of the following physical ideas is closely connected with the question of free will?
    * a. the uncertainty principle
      b. the exclusion principle
      c. the principle of complementarity
      d. wave/particle duality

50. The light from a laser is
      a. monochromatic.
      b. in phase.
      c. coherent.
    * d. All of the above are correct.

51. Two coherent sources of light shine on a distant screen. If we want to calculate the intensity of the light at positions on the screen, we
    * a. add the displacements and then square the result.
      b. square the displacements and then add the result.
      c. add the intensities and then square the result.
      d. square the intensities and then add the result.

52. Which of the following processes is crucial to the operation of a laser?
      a. spontaneous emission
      b. spontaneous absorption
    * c. stimulated emission
      d. stimulated absorption

Chapter 25: The Nucleus

1. When Becquerel was trying to produce X rays by shining light on fluorescent materials, he discovered nuclear radiation. What observation convinced him that he was not simply observing X rays?
   - a. They penetrated materials.
   - b. They exposed photographic plates.
   - c. They ionized air molecules.
   - * d. There was no need for a cathode-ray tube.

2. If radium (which is radioactive) and chlorine (which is not radioactive) combine to form radium chloride, the compound is
   - a. no longer radioactive.
   - * b. as radioactive as the radium content.
   - c. half as radioactive.
   - d. twice as radioactive.

3. Which of the following factors affects the amount of radioactivity of a sample of radium?
   - a. external pressure
   - b. temperature
   - * c. amount of radium
   - d. presence of electric and magnetic fields

4. Alpha particles are now known to
   - a. have a charge of −2.
   - b. be electromagnetic radiation.
   - c. be electrons.
   - * d. be the nuclei of helium atoms.

5. Beta particles are now known to
   - a. have a charge of −2.
   - b. be electromagnetic radiation.
   - * c. be electrons.
   - d. be the nuclei of helium atoms

6. Three of the names listed below refer to the same thing. Which one does NOT?
   - a. beta particles
   - b. cathode rays
   - * c. gamma rays
   - d. electrons

Chapter 25   The Nucleus

7. Gamma rays are
      a. positively charged electrons.
      b. electrons.
      c. the nuclei of helium atoms.
  * d. high-energy photons.

8. The proton was discovered
      a. as a product of radioactive decay.
  * b. by bombarding light elements with alpha particles.
      c. during attempts to prove that water was a compound.
      d. in experiments using particle accelerators.

9. Nuclei are known to consist of
      a. protons and electrons.
      b. neutrons and electrons.
  * c. protons and neutrons.
      d. protons, neutrons, and electrons.

10. The size of the atomic mass unit is attained by setting the mass
      a. of the neutral oxygen atom equal to 8.
  * b. of the neutral carbon atom equal to 12.
      c. of the oxygen nucleus equal to 8.
      d. of the carbon nucleus equal to 12.

11. The atomic mass of a nucleus in atomic mass units is approximately given by the number of its
      a. protons.
      b. neutrons.
  * c. nucleons.
      d. electrons.

12. What is the approximate mass of $^{90}_{40}Zr$ in atomic mass units?
      a. 40
      b. 50
  * c. 90
      d. 130

13. The three naturally occurring isotopes of neon are $^{20}_{10}$Ne , $^{21}_{10}$Ne , and $^{22}_{10}$Ne . Given that the atomic mass of naturally occurring neon is 20.18 atomic mass units, which of these three isotopes must be the most common?

    *   a. $^{20}_{10}$Ne

           b. $^{21}_{10}$Ne

           c. $^{22}_{10}$Ne

           d. They are all equally abundant.

14. The chemical identity of an atom is determined by the number of _____ in its nucleus.

    *   a. protons
           b. neutrons
           c. electrons
           d. nucleons

15. Atoms whose nuclei contain the same number of protons but different numbers of neutrons are called

           a. radioactive.
           b. daughters.
    *   c. isotopes.
           d. nucleons.

16. How many protons are in the nucleus of $^{21}_{10}$Ne ?

    *   a. 10
           b. 11
           c. 21
           d. 31

17. How many neutrons are in the nucleus of $^{21}_{10}$Ne ?

           a. 10
    *   b. 11
           c. 21
           d. 31

18. How many nucleons are in the nucleus of $^{21}_{10}$Ne ?

           a. 10
           b. 11
    *   c. 21
           d. 31

19. How many protons are in the nucleus of $^{47}_{18}$Ar ?

    *   a. 18
        b. 29
        c. 47
        d. 65

20. How many neutrons are in the nucleus of $^{47}_{18}$Ar ?

        a. 18
    *   b. 29
        c. 47
        d. 65

21. How many nucleons are in the nucleus of $^{47}_{18}$Ar ?

        a. 18
        b. 29
    *   c. 47
        d. 65

22. Which of the following combinations of nucleons would NOT form an isotope of hydrogen?
        a. a single proton
        b. one proton and one neutron
    *   c. two protons and one neutron
        d. one proton and two neutrons

23. What daughter is formed when $^{239}_{94}$Pu decays by alpha decay?

        a. $^{235}_{92}$Pu
    *   b. $^{235}_{92}$U
        c. $^{239}_{93}$Np
        d. $^{239}_{94}$Pu

24. $^{240}_{94}$Pu can be formed by alpha decay. What is the parent isotope?

    *   a. $^{244}_{96}$Cm
        b. $^{244}_{96}$Pu
        c. $^{239}_{93}$Np
        d. $^{236}_{92}$U

25. What daughter is formed when $^{239}_{93}$Np decays by beta minus decay?

     a. $^{235}_{91}$Pa

     b. $^{239}_{96}$Pu

*   c. $^{239}_{94}$Pu

     d. $^{239}_{92}$Pu

26. What daughter is formed when $^{18}_{7}$N decays by beta minus decay?

*   a. $^{18}_{8}$O

     b. $^{18}_{6}$C

     c. $^{19}_{8}$O

     d. $^{17}_{6}$C

27. $^{128}_{49}$In can be formed by beta minus decay. What is the parent isotope?

     a. $^{128}_{50}$Sn

*   b. $^{128}_{48}$Cd

     c. $^{129}_{50}$Sn

     d. $^{127}_{48}$Cd

28. What daughter is formed when $^{15}_{8}$O decays by beta plus decay?

     a. $^{16}_{8}$O

     b. $^{15}_{9}$Fl

*   c. $^{15}_{7}$N

     d. $^{11}_{6}$C

29. $^{50}_{24}$Cr can be formed by beta plus decay. What is the parent isotope?

     a. $^{50}_{23}$V

     b. $^{50}_{25}$Cr

*   c. $^{50}_{25}$Mn

     d. $^{51}_{25}$Mn

30. What daughter is formed when $^{7}_{4}$Be decays via electron capture?

      a. $^{6}_{4}$Be

*   b. $^{7}_{3}$Li

      c. $^{6}_{3}$Li

      d. $^{7}_{5}$B

31. $^{65}_{29}$Cu can be formed by electron capture decay. What is the parent isotope?

      a. $^{64}_{28}$Ni

      b. $^{65}_{28}$Ni

      c. $^{66}_{30}$Zn

*   d. $^{65}_{30}$Zn

32. If a nucleus decays by emitting an alpha particle, the daughter has
    a. one less charge and almost no change in mass.
    b. one less charge and one amu less mass.
    c. two fewer charges and two amu less mass.
  * d. two fewer charges and four amu less mass.

33. If a nucleus decays by emitting a beta minus, the daughter has
    a. one less charge and almost no change in mass.
    b. one less charge and one amu less mass.
  * c. one more charge and almost no change in mass.
    d. one more charge and one amu less mass.

34. If a nucleus decays via electron capture, the daughter has
  * a. one less charge and almost no change in mass.
    b. one less charge and one amu less mass.
    c. one more charge and almost no change in mass.
    d. one more charge and one amu less mass.

35. If a nucleus decays by emitting a beta plus, the daughter has
  * a. one less charge and almost no change in mass.
    b. one less charge and one amu less mass.
    c. one more charge and almost no change in mass.
    d. one more charge and one amu less mass.

36. What type of decay process is involved in the reaction $^{228}_{92}U \rightarrow ^{228}_{91}Pa + (?)$?

        a. alpha
*   b. beta plus
        c. beta minus
        d. electron capture

37. What type of decay process is involved in the reaction $^{254}_{100}Fm \rightarrow ^{250}_{98}Cf + (?)$?

*   a. alpha
        b. beta plus
        c. beta minus
        d. electron capture

38. What type of decay process is involved in the reaction $^{233}_{90}Th \rightarrow ^{233}_{91}Pa + (?)$?

        a. alpha
        b. beta plus
*   c. beta minus
        d. electron capture

39. What does the (?) stand for in the reaction $^{18}_{9}F \rightarrow ^{17}_{8}O + (?)$?

*   a. proton
        b. alpha
        c. beta minus
        d. beta plus

40. A proton strikes a nucleus of $^{20}_{10}Ne$. Assuming an alpha particle comes out, what isotope is produced?

        a. $^{14}_{6}C$
        b. $^{15}_{6}C$
*   c. $^{17}_{9}F$
        d. $^{16}_{8}O$

41. A nucleus of $^{14}_{7}N$ can absorb a neutron and form a new stable nucleus of

        a. $^{14}_{6}C$
        b. $^{15}_{6}C$
*   c. $^{15}_{7}N$
        d. $^{14}_{8}O$

42. A nucleus of $^{238}_{92}$U captures a neutron to form an unstable nucleus that undergoes two successive beta minus decays. The resulting nucleus is
    a. $^{239}_{92}$U .
    b. $^{239}_{93}$Np .
    c. $^{238}_{94}$Pu .
    * d. $^{239}_{94}$Pu .

43. A radioactive material has a half-life of 75 days. How long would you have to watch a particular nucleus before you would see it decay?
    a. 75 days
    b. 150 days
    c. an infinite length of time
    * d. There is no way of knowing.

44. A radioactive material has a half-life of 3 minutes. If you begin with 512,000 radioactive atoms, approximately how many would you expect to have after 12 minutes?
    a. 256,000
    b. 128,000
    c. 64,000
    * d. 32,000

45. If the activity of a particular radioactive sample is 40 Ci and its half-life is 100 years, in 200 years we would expect to have an activity of approximately
    a. 20 Ci.
    * b. 10 Ci.
    c. 5 Ci.
    d. zero.

46. If a gram of radium has an activity of 1 curie, what is the activity of 4 grams of radium?
    a. 0.25 curie
    b. 1 curie
    * c. 4 curie
    d. 16 curie

47. Radioactive clocks are useful for determining the age of things that were once living because
    * a. the rate of radioactive decay is unaffected by environmental conditions.
    b. radioactive decay is faster than biological decay.
    c. radioactive decay is slower than biological decay.
    d. radioactive clocks are not slowed by time dilation.

48. The age of Earth has been determined by using radioactive clocks to be approximately _____ years old.
    - a. 6000 years
    - b. 100 thousand years
    - c. 4.5 million years
    - * d. 4.5 billion years

49. If the ratio of C-14 to C-12 in a piece of parchment is only 1/16 of the atmospheric ratio, how old is the parchment? The half-life of C-14 is 5700 years.
    - a. 11,400 years
    - b. 17,100 years
    - * c. 22,800 years
    - d. 45,600 years

50. As alpha and beta particles pass through matter, they interact with the matter primarily through
    - a. collisions with nuclei.
    - * b. the ionization of atoms.
    - c. the excitation of electrons into higher orbits.
    - d. diffraction.

51. A beam containing electrons, protons, and neutrons is aimed at a concrete wall. If all particles have the same kinetic energy, the _____ will penetrate the farthest into the wall, while the _____ will hardly penetrate at all.
    - a. electrons ... protons
    - b. protons ... electrons
    - c. neutrons ... electrons
    - * d. neutrons ... protons

52. Which of the following particles interacts with matter in a way that is fundamentally different from the interactions of the other particles?
    - a. protons
    - b. neutrons
    - * c. gamma rays
    - d. alpha particles

53. By what factor is the intensity of a beam of gamma rays reduced after passing through 3 half-distances of material?
    - a. 3
    - b. 6
    - * c. 8
    - d. 9

54. Which of the following is a unit that reflects the biological effect caused by radiation?
      a. rad
  * b. rem
      c. joule
      d. curie

55. Which of the following is a unit that reflects the amount of energy deposited in a material by radiation?
  * a. rad
      b. rem
      c. epa
      d. curie

56. Which of the following sources of radiation contributes the least to the average yearly dose received by humans?
      a. cosmic rays
      b. internal decays within the human body
  * c. nuclear power
      d. medical

57. Which of the following sources of radiation contributes the most to the average yearly dose received by humans?
      a. cosmic rays
      b. internal decays within the human body
      c. nuclear power
  * d. medical

Chapter 26: Nuclear Energy

1. An electron, a positron, and a proton are each accelerated through a potential difference of one million volts. Which of these acquires the largest kinetic energy?
   - a. electron
   - b. positron
   - c. proton
   - * d. They acquire equal kinetic energies.

2. A positron, a proton, and an alpha particle are each accelerated through a potential difference of one million volts. Which of these acquires the largest kinetic energy?
   - a. positron
   - b. proton
   - * c. alpha particle
   - d. They acquire equal kinetic energies.

3. A positron, a proton, and an alpha particle are each accelerated through a potential difference of one million volts. Which of these acquires the largest momentum?
   - a. positron
   - b. proton
   - * c. alpha particle
   - d. They acquire equal momenta.

4. What potential difference is required to accelerate an alpha particle to a kinetic energy of 4 million electron volts?
   - * a. 2 million volts
   - b. 4 million volts
   - c. 8 million volts
   - d. 16 million volts

5. What electric potential difference is required to accelerate a proton to an energy of 5 million electron volts?
   - * a. 5 million volts
   - b. 5 million joules
   - c. 10 million volts
   - d. 10 million joules

6. What is the final kinetic energy of an alpha particle accelerated through a potential difference of 20 million volts?
   - a. 20 million volts
   - b. 20 million electron volts
   - c. 40 million volts
   - * d. 40 million electron volts

7. A proton and an alpha particle are each accelerated through the same potential difference. Which particle has the greater final speed?
   *   a. proton
       b. alpha particle
       c. They both will have the same final speed.
       d. There is not enough information to say.

8. The main advantage of a linear accelerator over a circular accelerator is that the linear accelerators
       a. require less land for construction.
   *   b. have smaller energy losses due to acceleration.
       c. are better suited for accelerating particles with larger masses.
       d. are simpler to understand.

9. Which of the following does NOT state a difference between the electric and strong forces?
       a. The electric force has an infinite range, while the strong force has a very short range.
       b. The strong force is approximately 100 times stronger.
   *   c. The strong force is only attractive, while the electric force can be attractive or repulsive.
       d. The strength of the strong force depends on more than the distance between the two particles.

10. Which of the following forces is responsible for holding the nucleons together to form nuclei?
        a. electric
        b. gravitational
    *   c. strong
        d. weak

11. How do the forces between two protons (p-p), two neutrons (n-n), and a neutron and a proton (n-p) compare?
    *   a. They are all virtually the same.
        b. The p-p force is the strongest.
        c. The p-p force is the weakest.
        d. The n-n force is the weakest.

12. Which of the following forces is responsible for the beta decay processes?
        a. gravitational
        b. electromagnetic
        c. strong
    *   d. weak

13. Which of the following is NOT one of the four forces in nature?
    *   a. inertial
        b. strong
        c. weak
        d. electromagnetic

14. Which of the following lists the forces in the order of decreasing strength?
    *   a. strong, electromagnetic, weak, gravitation
        b. strong, weak, electromagnetic, gravitation
        c. strong, electromagnetic, gravitation, weak
        d. electromagnetic, strong, gravitation, weak

15. Which of the following is released when a neutron and a proton combine to form a deuteron?
        a. neutron
        b. proton
        c. heat
    *   d. gamma ray

16. Which of the following has the largest mass?
    *   a. 90 protons and 150 neutrons
        b. $^{110}_{40}Zr + {}^{130}_{50}Sn$
        c. $^{240}_{90}Th$
        d. $^{120}_{45}Rh + {}^{120}_{45}Rh$

17. Which of the following has the smallest mass?
        a. 96 protons and 138 neutrons
        b. $^{234}_{96}Cm$
    *   c. $^{110}_{46}Pd + {}^{124}_{50}Sn$
        d. All three have the same mass.

18. Both $^{12}_{7}N$ and $^{12}_{5}B$ decay naturally to the stable nucleus $^{12}_{6}C$. Which of these three nuclei has the smallest mass?
        a. N-12
        b. B-12
    *   c. C-12
        d. They have equal masses.

19. $^{14}_{6}C$ decays to $^{14}_{7}N$ via beta minus decay. Which nucleus has the larger mass?

   *   a. $^{14}_{6}C$

        b. $^{14}_{7}N$

        c. They have the same mass.

20. $^{17}_{7}N$ beta decays to $^{17}_{8}O$ with a reaction energy of 8.68 MeV. $^{17}_{9}F$ beta-plus decays to $^{17}_{8}O$ with a reaction energy of 2.76 MeV. Which parent nucleus has the greater mass?

   *   a. $^{17}_{7}N$

        b. $^{17}_{9}F$

        c. They have the same mass.

21. Which nuclei have the largest average binding energies per nucleon? The ones with nucleon numbers
        a. less than 20
   *   b. between 30 and 100
        c. more than 200
        d. They are all the same.

22. Which nucleus would have the greater total binding energy, $^{56}_{26}Fe$ or $^{112}_{48}Cd$?

        a. $^{56}_{26}Fe$

   *   b. $^{112}_{48}Cd$

        c. They would have the same total binding energy.

23. Which nucleus would have the greater binding energy per nucleon, $^{56}_{26}Fe$ or $^{112}_{48}Cd$?

   *   a. $^{56}_{26}Fe$

        b. $^{112}_{48}Cd$

        c. They would have the same total binding energy.

24. Most stable nuclei with small atomic numbers have _____, while those with large atomic numbers have _____.
        a. more neutrons than protons ... more protons than neutrons
        b. more protons than neutrons ... more neutrons than protons
   *   c. equal numbers of neutrons and protons ... more neutrons than protons
        d. equal numbers of neutrons and protons ... more protons than neutrons

25. How would you expect an unstable nucleus of $_{30}^{80}$Zn to decay?

      a. alpha particle
      b. beta plus
  * c. beta minus
      d. electron capture

26. What is the most likely decay mode for $_{11}^{20}$Na ?

      a. alpha particle
  * b. beta plus
      c. beta minus
      d. electron capture

27. Nuclear fission
  * a. is the splitting of heavy nuclei into lighter ones.
      b. is the combining of light nuclei to form heavier ones.
      c. results from a series of alpha decays.
      d. releases much less energy per atom than a chemical process.

28. Assume a $_{92}^{235}$U nucleus absorbs a neutron and fissions with the release of three neutrons. If one of the fission fragments is $_{56}^{144}$Ba , how many protons and neutrons are there in the other fission fragment?

      a. 33 protons and 56 neutrons
      b. 36 protons and 56 neutrons
  * c. 36 protons and 53 neutrons
      d. 39 protons and 53 neutrons

29. How many neutrons are released in the following fission reaction?
$$_{0}^{1}n + _{92}^{235}U \rightarrow _{54}^{140}Xe + _{38}^{94}Sr + (?)\,_{0}^{1}n$$

      a. 1
  * b. 2
      c. 3
      d. 4

30. The fissioning of ^{235}U can produce a chain reaction because it
      a. emits photons.
      b. is unstable to radioactive decay.
  * c. releases two or three neutrons.
      d. converts mass into energy.

Chapter 26   Nuclear Energy

31. Which of the following does NOT determine whether a piece of uranium undergoes a subcritical or supercritical reaction?
    a. mass
    b. volume
    c. ratio of isotopes
    * d. temperature

32. Which of the following is NOT needed in a fission reactor?
    a. a moderator
    b. neutron absorbing control rods
    c. a critical mass of fuel
    * d. a separate source of neutrons

33. Why do most fission reactors employ moderators?
    a. to moderate the reactions so the reactor doesn't blow up
    b. to absorb excess neutrons to control the power output
    * c. to slow the neutrons down
    d. to increase the breeding of new fuel

34. Why does a fission reactor have control rods?
    a. to provide the extra neutrons needed to run the reactor
    * b. to absorb excess neutrons to control the power output
    c. to slow the neutrons down
    d. to increase the breeding of new fuel

35. In the first cycle of a fission chain reaction, a single nucleus fissions and produces three neutrons. If every free neutron initiates a new fission event, three nuclei fission in the second cycle, for a total of four. What is the total number of fission events after five cycles?
    a. 31
    b. 40
    * c. 121
    d. 256

36. In the first cycle of a fission chain reaction, a single nucleus fissions and produces two neutrons. If every free neutron initiates a new fission event, two nuclei fission in the second cycle, for a total of three. If each fission event releases 210 MeV on average, what is the total energy released after eight cycles?
    a. 2.69×10^4 MeV
    * b. 5.36×10^4 MeV
    c. 6.02×10^6 MeV
    d. 8.24×10^8 MeV

37. Breeder reactors
      a. are much safer to operate than boiling-water reactors.
      b. do not have the problem of producing bomb-grade materials.
 *   c. generate more fuel than they use.
      d. are an example of getting something for nothing.

38. Which of the following processes releases energy?
      a. fusion of heavy nuclei
      b. fission of intermediate nuclei
      c. fission of light nuclei
 *   d. fusion of light nuclei

39. Nuclear fusion
      a. is the splitting of heavy nuclei into lighter ones.
 *   b. is the combining of light nuclei to form heavier ones.
      c. results from a series of alpha decays.
      d. releases much less energy per atom than a chemical process.

40. Fusion reactors require very high temperatures
      a. because the electrical generators require very high temperatures.
 *   b. for the hydrogen nuclei to get close enough together to fuse.
      c. to cause the uranium nuclei to split when they collide.
      d. to minimize the production of radioactive waste.

41. The temperature of the plasma in a typical household fluorescent light is 20,000°C. Why can
    you touch an operating light without being burned?
      a. The tube is made from a special insulating glass.
 *   b. The density is very low, so the plasma contains little heat energy.
      c. The heat capacity of glass is very high.
      d. The plasma does not touch the glass tube.

42. Which process releases the most energy per nucleon?
      a. burning carbon
      b. fissioning of uranium
 *   c. fusing of hydrogen
      d. fissioning of iron

43. Where do we obtain the fuel for fusion reactors?
    a. in underground mines
    b. from the radioactive wastes of power reactors
*   c. from water
    d. from natural gas

44. The source of the Sun's energy is
    a. gravitational collapse.
    b. the burning of coal.
*   c. nuclear fusion.
    d. nuclear fission.

Chapter 27: Elementary Particles

1. Which of the following was NOT on the list of elementary particles before 1932?
   - a. photon
   - b. electron
   - * c. positron
   - d. proton

2. Which of the following particles on the list of elementary particles in 1932 are NOT on the current list?
   - a. photon
   - b. electron
   - * c. proton
   - d. positron

3. Which of the following is a particle-antiparticle pair?
   - a. proton -- positron
   - b. proton -- neutron
   - c. neutron -- neutrino
   - * d. electron -- positron

4. Which of the following is the antiparticle of the electron?
   - a. photon
   - b. proton
   - * c. positron
   - d. neutrino

5. If we ignore the sign, which of the following properties of particles and their corresponding antiparticle do NOT have the same size?
   - a. mass
   - b. charge
   - c. spin
   - * d. All of these have the same size.

6. What is the ultimate fate of an antiparticle here on Earth?
   - a. As soon as it encounters any matter, it is annihilated.
   - * b. It annihilates only with its corresponding particle.
   - c. It spontaneously turns into energy according to $E = mc^2$.
   - d. It normally forms an antiatom.

7. Which of the following particles could be used to communicate with an antiworld?
*   a. photon
    b. electron
    c. proton
    d. neutron

8. If an antiproton and an antineutron combined to form an antideuteron, the amount of energy released is equal to
    a. the total rest mass of the antiproton and the antineutron.
    b. the rest mass of the antideuteron.
*   c. the amount released in the formation of a deuteron.
    d. zero.

9. When an electron annihilates with a positron, the amount of energy released is equal to the
*   a. total rest mass of the electron and the positron.
    b. rest mass of the electron.
    c. rest mass of the positron.
    d. binding energy of the hydrogen atom.

10. The mass of a positron is
*   a. the same as that of an electron.
    b. the negative of that of an electron.
    c. zero.
    d. about the same as that of a neutron.

11. The mass of a photon is
    a. the same as that of an electron.
    b. the negative of that of an electron.
*   c. zero.
    d. about the same as that of a neutron.

12. What is the antiparticle of the photon?
    a. tau neutrino
    b. intermediate vector boson
    c. positron
*   d. It is its own antiparticle.

13. The rest mass of the positron is 0.511 MeV. Why is more than 0.511 MeV released when a positron is annihilated?
   a. Energy is not conserved in the annihilation.
 * b. An electron is also annihilated.
   c. The kinetic energy of the positron accounts for all of the excess energy.
   d. The positron is usually in an excited state.

14. The initial observations of beta decay indicated that most of the classical conservation laws might be violated. Which of the following was obeyed without inventing the neutrino?
   a. energy
 * b. charge
   c. angular momentum
   d. linear momentum

15. Observations of beta decay showed that electrons are emitted with a
   a. fixed kinetic energy.
   b. fixed momentum.
 * c. range of kinetic energies.
   d. range of masses.

16. Why did it take so long for experimentalists to detect the neutrino?
   a. It has no mass.
   b. It has no charge.
   c. It does not live very long.
 * d. It only interacts via the weak interaction.

17. The neutrino interacts with the world primarily through the _____ force.
   a. strong
   b. electromagnetic
 * c. weak
   d. gravitational

18. The conservation laws of energy and momentum
   a. are valid for all situations at all levels.
   b. are always obeyed at the macroscopic level, but always violated at the atomic level.
 * c. may be violated at the atomic level if the violation does not last too long.
   d. are no longer valid at any level.

19. What is the exchange particle for the electromagnetic interaction?
   a. intermediate vector boson
   b. electron
   c. muon
  * d. photon

20. What feature of the exchange particle accounts for the infinite range of the electromagnetic interaction?
   a. It has no charge.
   b. It satisfies conservation of energy and momentum at all times.
  * c. It has no rest mass.
   d. It is its own antiparticle.

21. Which of the following is NOT a particle that is exchanged to produce a force between two other particles?
  * a. neutron
   b. pion
   c. graviton
   d. intermediate vector boson

22. What feature of the strong interaction required the exchange particles to have non-zero masses?
   a. strength
  * b. short range
   c. orientation
   d. polarization

23. Which argument can be used to support the idea that the exchange particle for the gravitational interaction has zero rest mass? The gravitational force
   a. is the weakest of the four forces.
   b. does not depend on charge.
  * c. has an infinite range.
   d. is proportional to the masses of the objects.

24. Which of the following do NOT participate in the strong interaction?
   a. hadrons
  * b. leptons
   c. baryons
   d. mesons

25. Which one of the following is NOT a member of the lepton family?
    a. electron
    b. muon
 *   c. proton
    d. neutrino

26. Which of the following does NOT have a spin of 1/2?
    a. proton
    b. electron
 *   c. pion
    d. neutrino

27. If a particle has a life-time of approximately 10^{-10} seconds, via what interaction does it decay?
    a. strong
    b. electromagnetic
 *   c. weak
    d. gravitational

28. If a particle has a life-time of approximately 10^{-16} seconds, via what interaction does it decay?
    a. strong
 *   b. electromagnetic
    c. weak
    d. gravitational

29. Which of the following quantities is conserved in strong or electromagnetic interactions but violated in weak interactions?
    a. energy
    b. charge
    c. baryon number
 *   d. strangeness

30. The existence of intermediate vector bosons was predicted many years before they were detected in an accelerator. Why did this discovery take so long when scientists knew what they were looking for?
    a. They have no charge.
    b. They only interact via the weak interaction.
 *   c. They have very large masses.
    d. They have charm.

Chapter 27   Elementary Particles

31. What conservation law forbids the following reaction from occurring?

$$\pi^- + p \rightarrow \pi^+ + \bar{p}$$

    a. charge
    b. strangeness
 *   c. baryon number
    d. spin

32. What conservation law forbids the decay of the proton according to the following process?
$$p \rightarrow \pi^o + e^+$$

    a. charge
    b. energy
    c. spin
 *   d. baryon number

33. What is the X in the pion decay $\pi^+ \rightarrow \mu^+ + X$?
    a. proton
    b. neutron
    c. electron
 *   d. mu neutrino

34. What is the X in the antimuon decay $\mu^+ \rightarrow \nu_e + \bar{\nu}_\mu + X$?
    a. electron
 *   b. positron
    c. photon
    d. neutron

35. Quarks
    a. were discovered in a bubble chamber experiment.
 *   b. combine to form the baryons and mesons.
    c. combine to form the leptons.
    d. have integral multiples of the charge on the electron.

36. A neutron consists of two down quarks and one up quark. What particle is made up of two down antiquarks and one up antiquark?
    a. neutrino
    b. proton
    c. antiproton
 *   d. antineutron

37. A proton consists of two up quarks and one down quark. What particle is made up of two up antiquarks and one down antiquark?
    a. neutrino
    b. neutron
 * c. antiproton
    d. antineutron

38. Which of the following is NOT considered to be elementary; that is, which one is a composite of other particles?
    a. neutrino
    b. muon
 * c. neutron
    d. quark

39. Given the following properties of the up and down quarks and of the negative pion, what combination of quarks makes up a negative pion?
    up: charge = +2/3; baryon = 1/3; spin = 1/2
    down: charge = −1/3; baryon = 1/3; spin = 1/2
    pi minus: charge = −1;   baryon = 0,   spin = 0
    a. ud
    b. ūd
    c. ud̄
 * d. ūd

40. A particle consists of a top quark and its corresponding antiquark. This particle is a
 * a. meson.
    b. baryon.
    c. lepton.
    d. graviton.

41. Given the following properties of the up and down quarks and of the positive pion, what combination of quarks makes up a positive pion?
    up: charge = +2/3; baryon = 1/3; spin = 1/2
    down: charge = −1/3; baryon = 1/3; spin = 1/2
    pi plus: charge = +1;   baryon = 0,   spin = 0
    a. ud
    b. ūd
 * c. ud̄
    d. ūd

42. Given the following properties of the up and down quarks and of the proton, what combination of quarks makes up a proton?
    up: charge = +2/3; baryon = 1/3; spin = 1/2
    down: charge = −1/3; baryon = 1/3; spin = 1/2
    proton: charge = +1;   baryon = 1,   spin = 1/2
        a. uuu
    *   b. uud
        c. udd
        d. ddd

43. What charge would a baryon have if it was composed of a down quark, a strange quark, and a top quark?
        a. -1
    *   b. zero
        c. +1
        d. +3

44. Which of the following is NOT a flavor of quark?
        a. strange
        b. charm
        c. top
    *   d. red

45. Color has been proposed as an additional attribute of quarks to
        a. make sure that we can distinguish between neutrons and protons.
        b. solve the problem of beta decay.
        c. make sure there are enough quarks to match the leptons.
    *   d. solve the problem of the omega minus and the exclusion principle.

46. In the original quark model, the Ω^- was believed to be composed of three strange quarks. This assumption causes problems because quarks are expected to obey the Pauli exclusion principle. How did physicists get around this problem?
        a. They discarded the Pauli exclusion principle.
    *   b. They postulated the existence of another quantum number called color.
        c. They invoked Heisenberg's uncertainty principle to account for the violation.
        d. They postulated that the Pauli exclusion principle does not apply to strange particles.

47. What exchange particle holds the quarks together?
    a. photons
    b. mesons
 * c. gluons
    d. hadrons

48. Which of the following is NOT a flavor of color charge?
    a. red
 * b. yellow
    c. blue
    d. green

Chapter 28: Frontiers

1. Which of the four fundamental forces is not included in the current grand unified theories?
    a. strong
    b. electromagnetic
    c. weak
    * d. gravitational

2. Which of the following is NOT a possible source of gravitational waves?
    a. supernovae
    b. orbiting neutron stars
    * c. a stationary black hole
    d. orbiting black holes

3. Why are gravity waves so hard to detect?
    * a. They are 10^{43} times weaker than electromagnetic waves.
    b. Gravitons have no charge.
    c. Gravitons have no mass.
    d. They are polarized.

4. If orbiting planets are constantly radiating energy in the form of gravitational waves, why don't the planets spiral into the Sun?
    a. The orbits are quantized.
    b. The centrifugal force holds them out.
    * c. The energy radiated in the form of gravitational waves is very small.
    d. The centripetal force holds them out.

5. The Planck model accounts for the stability of atoms by asserting that atoms only radiate when an electron jumps from one allowed orbit to another allowed orbit. Planets radiate gravitational waves
    * a. continually.
    b. only when they change orbits.
    c. only if their orbit is highly elliptical.
    d. only when they explode.

6. What indirect evidence has been found to support the existence of gravitational waves?
    a. Congress has funded the LISA project.
    * b. The orbital period of a binary system of neutron stars was found to be decreasing.
    c. Three orbiting satellites were found to appear closer together after the explosion of Supernova 1987A.
    d. The standard meter stick in France changed length briefly in 1998.

7. Neutron stars are a little more massive than our Sun. Why would a binary system of neutron stars be a strong source of gravitational waves?
    a. Neutron stars are much more dense than our Sun.
    * b. They have small orbits, requiring very large accelerations.
    c. Neutrons emit gravitons.
    d. Their orbital periods tend to be very large.

8. Superstring theory tries to include which of the basic forces with the others?
    a. electromagnetic
    b. weak
    * c. gravitational
    d. strong

9. The fundamental building blocks in current superstring theories have a characteristic size of
    * a. 10^{-35} m
    b. 10^{-15} m
    c. 10^{-10} m
    d. 10^{-4} m

10. What is the current estimate of the age of the Universe?
    a. 6000 years
    b. 1 billion years
    * c. 14 billion years
    d. 20 billion years

11. There was a period of time right after the Big Bang when the laws of physics as we know them did not exist and it was not necessary to take college physics classes. How long did this period last?
    * a. 10^{-43} second
    b. 0.01 second
    c. 300,000 years
    d. 4 million years

12. Which of the fundamental forces was the first to become distinct from the others after the Big Bang?
    a. electromagnetic
    b. weak
    * c. gravitational
    d. strong

13. How long did it take after the Big Bang before the four fundamental forces of nature were distinct from each other?
      a. 10^{-43} second
  * b. 10^{-6} second
      c. 300,000 years
      d. 4 million years

14. Why do neutrino experiments have to be performed in mineshafts deep below Earth's surface?
  * a. Cosmic radiation swamps the neutrino signals on the surface.
      b. The experiments require controlled temperature.
      c. The dark shafts eliminate the photon signals.
      d. The detectors involve toxic chemicals.

15. The fusion reactions in the Sun are now well understood. These reactions predict a much higher ratio of electron neutrinos compared to the other two flavors than is observed here on Earth. How do scientists account for this discrepancy?
      a. Electron neutrinos are much more likely to meet their antiparticle and annihilate.
      b. Electron neutrinos have a shorter lifetime than the other two flavors.
      c. Electron neutrinos are believed to be trapped by the Van Allen belt.
  * d. A neutrino is a superposition of all three flavors, and can oscillate from one state to another.

16. What evidence led scientists to believe that the Universe is flat?
      a. Columbus day is not celebrated in much of the world.
  * b. measurements of ripples in the cosmic ray background
      c. The Hubble constant was found to be changing.
      d. The Hubble telescope discovered the "edge" of the Universe.

17. What observation led scientists to conclude that the rate of expansion of the Universe is increasing?
  * a. Radiation from distant supernovae are dimmer than would be expected.
      b. The orbital period of a binary system of neutron stars was found to be decreasing.
      c. The Hubble telescope discovered the "edge" of the Universe.
      d. Ripples in the cosmic ray background are getting bigger.

18. The reductionist view asserts that
      a. the size of the universe is decreasing.
      b. any physical system, no matter how complex, may be reduced to a binary system.
      c. the number of elementary particles is small.
  * d. any physical system, no matter how complex, may be understood in terms of its
         component parts.